I'm an Earth Angel!

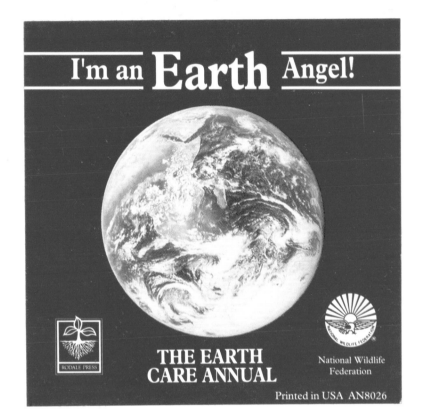

RODALE PRESS

THE EARTH CARE ANNUAL

National Wildlife
Federation

Printed in USA AN8026

THE EARTH CARE ANNUAL 1991

Edited by Russell Wild

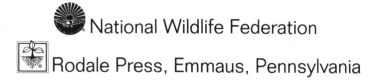

National Wildlife Federation

Rodale Press, Emmaus, Pennsylvania

In Memoriam

On September 20, 1990, just as this book went to press, Robert Rodale died in an automobile accident. Once, not long ago, an employee at Rodale Press asked the chairman what it was that he most wanted to communicate to the public. "I can do that in two words," he answered. "You Can!" This book carries the spirit of Robert Rodale, who wanted above all else a cleaner and healthier planet and had absolute faith that humankind could make it happen. Let us prove him right. Rest in peace, Bob.

The Editor

The Earth Care Annual **Team:**

Editor: *Russell Wild*

Research Coordinator: *Melissa Meyers*

Production Editor: *Jane Sherman*

Book Designer: *Lisa Farkas*

Cover Photo: *W. Perry Conway/*
   TOM STACK & ASSOCIATES

Reprint Permissions Coordinator: *Sandra
   Lloyd*

Copy Editor: *Sally Roth*

Indexer: *Lynne Hobbs*

Office Staff: *Eve Buchay, Karen
   Earl-Braymer, Roberta Mulliner*

Executive Editor, Health: *Carol Keough*

Vice President and Editor in Chief, Rodale
   Books: *William Gottlieb*

Other contributors to the *Earth Care Annual*
   project: *Judy Bridges, Anne Castaldo,
   Lynn Donches, Janet Glassman,
   Bill Howard, Bruce Mertz,
   Anthony Rodale, Maria Rodale,
   Bob Strohm, Mark Wexler, Lenore Wild*

Printed in the United States of America on
acid-free, recycled paper ∞

If you have any questions or comments
concerning this book, please write:

Rodale Press
Book Reader Service
33 East Minor Street
Emmaus, PA 18098

ISBN 0–87857–912–5 hardcover

2 4 6 8 10 9 7 5 3 1 hardcover

**Distributed in the book trade by St. Martin's
Press**

Contents

Earth Care for a New Decade

GARBAGE

GREENHOUSE EFFECT

OCEANS

OZONE

PESTICIDES

POLAR REGIONS

TOXIC WASTE

TROPICAL FORESTS

WATERWAYS

WILDLIFE

EARTH DAY PERSPECTIVE

Resources

ENVIRONMENTAL SHOPPER'S GUIDE

DIRECTORY OF ENVIRONMENTAL GROUPS

FOREWORD

By Jay D. Hair
President and Chief Executive Officer
National Wildlife Federation

Care for the land so it may give sustenance even unto the seventh generation. This is the Indian admonition given us by those who first called this continent home.

The idea: Our personal responsibility is to protect the land and water from which all life comes. The ethic is to live in harmony with nature, not to be its dominating master.

As far as our culture has traveled in generations of expansion and technology, that ethic still tugs at our conscience, because we know it is right. We are the earth's stewards, not its masters.

How far have we come from the rolling grasslands, the ancient forests, the pure rivers and lakes—all of them bountiful with wildlife—that the Indian knew? This book will help provide you with the answers. The story is grim. The achievement of individuals determined to restore environmental sanity is inspiring.

The environmental ethic was globally reaffirmed on last year's twentieth anniversary of Earth Day. This *is* the decade of the environment. We are moving toward a broadened understanding of the environment as a matter of national and even global security. To meet the challenges, we must fashion our beliefs into the force of action. From our own individual actions to the acts of Congress, from the nation's city halls to the state legislatures, we must insist on the environmental priority. The task demands creativity and courage. It will exact sacrifices. The nature of the threat is so pervasive that nothing short of the achievement of an environmental agenda is acceptable.

On the first Earth Day our challenges were immediate. We saw our lakes and streams dying, our land being sterilized with poisons, our air choked with toxic gas. The problem appeared at hand, the solution within reach. Now, a generation later, we are learning of the drastic consequences that turn rain into acid, strip our atmosphere of a vital shield of ozone, and portend an epoch of planetary warming. As our resources are being stressed beyond their ability to meet even the minimum needs of a third of the world's people, global population growth continues at quantum intensity. Our problems remain immediate and at hand, but we now know they are global, too. The planet's vital signs are warning us of danger to life on Earth.

As we look to the last decade of this century, we are confronted by another prospect undreamed of a generation ago. We are entering an era when it is no longer considered naive to think of bending the metal of war into the implements of sustenance.

A rendezvous between the environmental threat to global security and the means to meet it may be this generation's destiny. As this process unfolds, the United States must set an example and lead the way.

The environmental decade will require a redefinition of values. The concept of security needs to be defined as the sustainability of life, not the ability to produce armaments capable of ending it. This redefinition will require a reallocation of substantial funds from military stockpiles to environmental needs. Let there be no illusions here. Those who advance this cause need to be prepared to have their very patriotism challenged by those who misapprehend the dictates of loyalty.

Some new thinking is also required in our understanding of economics. The actual cost of pollution and environmental destruction must be factored into the price of production. The concepts of maximizing profit as an end in itself and of unlimited resource exploitation need redefinition to a model of a sustainable economy.

The environmental decade also requires a broadening of alliances. America's

minorities and economically disadvantaged must be aggressively included in the process of change. We must recognize that pollution's first and often worst victims are the poor. Americans of color bear a disproportionate burden of past decades of environmental recklessness. The perception of the environmental movement as a preserve of the privileged must be overcome.

In the broadest context, we need to make ourselves aware, to become an environmentally literate society.

In each of these areas the National Wildlife Federation and its 5.8 million members and supporters intend to continue as an aggressive force for responsible change. The publication of this book in association with Rodale Press is an example of that action.

As we comprehend the magnitude and complexity of the environmental problems facing us, the perception too often takes hold that we are simply observers to a drama we are powerless to influence.

The Earth Care Annual 1991 reinforces a fundamental principle of environmentalism: Individual action counts. The problems are global in nature, but they do not render the individual helpless. This volume, a compilation of some of the best environmental reporting over the past year, recounts the work of dedicated individuals who have fought to end the wanton destruction of sea life, to preserve the ecological dynamos of our tropical forests, to reverse toxic assaults on human existence, to protect our planet. It is, in a sense, a road map to help you be part of the solution, not of the problem. The annual is intended to jar conventional thinking, to upset the apple cart of complacency. From it you may draw both a sense of the dimensions of the task facing us and insight into the regenerative spirit of those who forge their caring into action.

Each of us can make environmentally responsible decisions. In the marketplace the dollar can become a vote for recyclable, nonpolluting products. America's manufacturers must be convinced that their incentive for profit lies in meeting a public demand for goods that do not degrade the environment. In scores of ways, many of them noted in this volume, we may each take actions that together have the multiplier effect of prompting change. In taking these actions, we should seek ways to amplify our voices by involving others.

In the political arena, our votes can count directly in firmly establishing an environmental agenda. Scrutinize the candidates, or become one, to ensure the realization of environmental reform. For its part, the National Wildlife Federation intends to be a leader as environmental organizations become more actively involved in the public decision-making process.

The ancient admonition to care for the earth has for too long been drowned out by the machinery of growth. It is up to us to secure a future in which once again the grass may grow and the rivers run free.

PORTRAITS OF EARTH CARE IN ACTION

Perhaps you feel the same way I do, that we—all of us working for a cleaner and healthier planet—are part of a large and growing club. If you've never thought of yourself as a member—let me welcome you.

Ever since I was a young boy, I've seen myself on a mission to provide care for our planet and its inhabitants. My particular interests have been in agriculture, personal health, and publishing.

And the publishing of the *Earth Care Annual 1991* is something I'm happy to be a part of. I consider this book—and I hope you will, too—as a sort of annual report for our world-wide club. In it, you'll read about other "members" and their activities over the past year.

Right now, I'd like to briefly introduce two young members, my son, Anthony, and my daughter, Maria.

Both care as much as I do about the condition of our planet. But each has chosen a different route to show concern. Anthony is a photographer, who strives to capture the essence of earth care on film. Maria is a journalist, working here at Rodale Press to produce our own magazines, newsletters, and books.

On the following pages are several of Anthony's photos, with comments from Maria.

The environment has no boundaries. Across different cultures and continents species may change, politics may change, and plant and animal life may take on whole new characters—but people everywhere have one thing in common: the need for an environment that can support us.

We also share the basic responsibility of caring for the earth. All over the world, people are doing what is within their capacity to do in order to improve their life and environment. The articles in this book are a testament to the spirit that is growing, and inspiring us to create a better environment.

The following photos, taken by Anthony on his travels around the world, are also a testament of the human spirit in action—people living and caring for the earth in the best way they are able.

In the struggle toward a cleaner environment and a better world, we must remember our priorities, our goals: That people and animals and plants and mountains can exist together in harmony, supporting each other and providing each other with their basic needs for existence—including happiness.

7:30 A.M. in Ghana, Africa (above). Gathering firewood is a daily fact of life for millions of Africans. Although their foraging contributes to global deforestation, until new alternatives are developed they have no choice but to use the trees that are available to them. The planting of new strains of fast-growing and drought-resistant trees may ease the burden for those who still rely on wood to heat, cook, and build. And new, highly efficient stoves are also being developed. But meanwhile, global deforestation is occurring at an astounding rate, with even greater damage occurring from farmers, ranchers, and loggers in the rainforests of Central and South America—often trying to supply our demand for inexpensive meats and hardwoods.

Grinding millet in Senegal, Africa (opposite page, top). Because resources are more difficult to obtain and food more difficult to grow, meal preparations become a community affair in many African villages—everyone helps and everyone shares. But producing food without harming the earth is one of the oldest challenges. Much of Africa's vulnerability to drought and famine is due to centuries of poor farming and grazing practices.

The Rodale Institute, a nonprofit organization devoted to creating new sustainable agricultural and community systems, is working in Senegal on easing Africa's vulnerability. We call it "famine prevention," which involves building up the "immunity" of a village by planting trees, planting more diverse crops, and using available resources for fertilizer—so when famine and drought do hit, the village is strong enough to survive and thrive.

Dancing in Harlem, USA (opposite page, bottom). The city is just as much a part of the world's environment as the rainforest, the desert, or a field. And in all types of environments all over the world, people deal with hardship, scarcity, and social problems—but a celebratory spirit often prevails. Local festivals such as the Friendly Place Cultural Arts festival in Harlem distribute family planning information (and condoms), literacy learning opportunities, and environmental education information amidst art and music. Such events go a long way toward building a brighter future, especially for the children.

Sidewalk market in Kharkov, Soviet Union (above). As countries develop and expand economically, it's important that mistakes aren't transported from culture to culture. And if they are, that the true costs are analyzed in the process. Many developing countries are more eager to have high technology, luxuries, and new inventions than they are to worry about the environmental damage that comes with them.

But it is human nature to want to grow—to improve our existence. We have to look for partnerships in developing alternatives and solutions to problems in order to accommodate our growth. For instance, Rodale Press, in a joint venture with the Soviets, will be publishing the first farming magazine in the Soviet Union. *Novii Fermer* magazine will give Soviet farmers desperately needed information on sustainable and organic agriculture. This type of exchange of information and ideas is vital, because a better environment is part of a better life for everyone.

Neighborhood friends in Santa Ana, Costa Rica (opposite page). Happiness is defined differently all over the world, but some basic needs will always remain: food, clean water, health care, shelter, clean air, love. If we all do our part to live more simply, more efficiently, and more responsibly, the word "scarcity" will refer to pain and suffering, not to grain, water, or a safe place to raise our families.

Morning class in Nanjing, China (above). As custodian of the largest population on Earth, China has a special challenge. For thousands of years the Chinese have been farming the land to its fullest capacity, and they remain some of the healthiest people on the planet. But as changes rapidly occur, they have been relying more and more on chemical fertilizers. Skyscrapers dot the sky in Beijing and cars are rapidly taking the place of bicycles in the larger cities.

Progress cannot be denied to any country, but the environmental impact of growth affects all of us. Once again, it's the children who will probably end up having to manage the damage and find creative solutions (one more reason to make sure education is a high priority)—unless we all act quickly to change our inefficient lifestyle to one that's more gentle to the earth.

I hope that you found these photos inspiring. And I trust that you will find the stories that follow equally inspiring. I think they illustrate the vast number of ways that you and members of your family can show care for the earth.

Until next year, keep up the good work, don't let the seriousness of our environmental problems make you feel hopeless, and—perhaps above all—recruit as many new members as you can into our "club."

Cordially,

Bob Rodale

EARTH CARE
FOR A NEW DECADE

G A R B A G E

Unlike other haulers, Earthworm Inc., of Boston, collects scrap paper during the day so it can answer people's questions. Shown aboard a load of paper in their truck are Robin Ingenthron (left) and Earthworm president Jeff Coyne. See story on page 18. (Neal J. Menschel/*Christian Science Monitor*)

BURIED ALIVE

By Melinda Beck
(with *Newsweek* bureau reports)

At a distance, the yellow, granulated mounds rising 250 feet over Staten Island might be mistaken for sand dunes—if not for the stench. Fresh Kills is the largest city dump in the world, home to most of Gotham's garbage, 24,000 tons each day brought around the clock by 22 barges. "We get it all here—your plastics, your Styrofoams, even stoves and refrigerators," says supervisor William Aguirre. "It was a valley when it started [in 1948]. Now it's a mountain." By the year 2000, Fresh Kills will tower half again as high as the Statue of Liberty and fill more cubic feet than the largest Great Pyramid of Egypt—provided it lives that long. The state of New York claims Fresh Kills is leaching 2 million gallons of contaminated gunk into the groundwater each day and has threatened to close the dump down in 1991. As a result, city sanitation commissioner Brendan Sexton left this message for newly elected Mayor David Dinkins: "Hi. Welcome to City Hall. By the way, you have no place to put the trash."

Neither do many American communities, from Philadelphia to Berkeley, Minneapolis to Jacksonville. More than two-thirds of the nation's landfills have closed since the late 1970s; one-third of those remaining will be full in the next five years. Federal law prohibits dumping trash into the ocean. Incineration is under attack on economic and environmental grounds. Recycling is gaining popularity, but currently only 11 percent of U.S. solid waste lives again as something else. And still the volume of garbage keeps growing—up by 80 percent since 1960, it's expected to mount an additional 20 percent by 2000. Not including sludge and construction wastes, Americans toss out 3.5 pounds of trash apiece each day, collectively 160 million tons each year. That's enough to spread 30 stories high over 1,000 football fields, enough to fill a bumper-to-bumper convoy of garbage trucks halfway to the moon.

Municipal waste haulers aren't taking it there—yet. But as their own landfills close and disposal fees soar, many com-

1

munities are trucking their trash across state lines and into rural areas. Some 28,000 tons of garbage travel the nation's highways each day; New York, Pennsylvania, and New Jersey export 8 million tons a year. That practice can be costly, too: Long Island townships each spend an average of $23 million a year shipping garbage out of state. Increasingly, it incenses residents on the receiving end as well. "I'm all for taking care of our garbage needs, but New Jersey can take care of its own," says state Sen. Roger Bedford of Alabama, which has placed a two-year moratorium on accepting garbage not grown at home.

NIMBY (Not in My Backyard) syndrome is breaking out all over, frustrating efforts to build new landfills, expand old ones, and site incinerators, transfer stations, and recycling centers. NIMBY has been joined by other acrimonious acronyms of the waste wars: GOOMBY (Get Out of My Backyard), LULU (Locally Undesirable Land Use), and NIMEY (Not in My Election Year). Greater Los Angeles has suffered them all in recent years. Angry citizens have three times blockaded the entrance to Lopez Canyon Landfill, hoping to close the last remaining city-owned dump. Officials of surrounding L.A. County say their canyons won't accept any more of the city's trash, and Mayor Tom Bradley helped scuttle the proposed $235 million Lancer incinerator when political opposition to the project grew ferocious. As a result, "we actually have garbage trucks running around town every day without a place to dump," says local environmentalist Will Baca.

That doesn't faze some unscrupulous drivers. As legal disposal grows more diffi-cult, some private waste haulers simply unload their fetid cargo anywhere, from ghetto streets to forests. Even the Mafia is concerned about the lack of landfill space. Law-enforcement officials say that two New York mob families, which own carting companies, are trying to gain control of valuable Pennsylvania dumps. Worse still, some truckers who haul meat and produce to the East in refrigerated vehicles are carrying maggot-infested garbage back West in the same trucks. Congress is considering banning the practice, which carries serious health risks. "Would you serve potato salad from your cat's litter box?" asked Pennsylvania State University food science professor Manfred Kroger at congressional hearings in August [1989].

Junk Mail

The garbage crisis didn't appear overnight, of course. Environmentalists first warned of it in the 1970s, and some citizens conscientiously toted cans, bottles, and paper to ragtag recycling centers. But there were scant markets for the recycled material and enthusiasm faded like last year's newsprint. The urgency seemed to wane as well: Garbage, after all, isn't as frightening as toxic waste or as photogenic as the burning Amazon. Meanwhile, the throwaway society has grown ever more disposable, substituting squeezable plastic ketchup bottles for glass, generating 12.4 billion glossy mail-order catalogs each year, and annually buying some 1 billion individual foil-lined boxes of fruit juice, complete with shrink wrapping and a plastic-encased straw on the side.

In the spring of 1987, Islip's wandering

garbage barge, laden with 3,000 tons of Long Island filth, was turned away from ports as far away as Belize. Its saga may have been to the trash crisis what the sinking of the *Lusitania* was to World War I, as nightly news stories starkly reminded Americans that what they toss out must go *somewhere*. Since then, 18 states and scores of municipalities have embarked on ambitious waste-reduction programs. Next July [1990], Minneapolis and St. Paul will ban all plastic food packaging that won't degrade or can't be recycled; Nebraska will ban most disposable diapers in 1993. With amazing speed, recycling has shed its tie-dyed image and attracted big-business investment and political passion. "Nobody knew what the heck curbside recycling was two years ago," says Gary Mielke of Illinois's Department of Energy and Natural Resources; now 500,000 households in his state alone set their glass, paper, and aluminum on the street in separate containers. The efforts seem to provide an outlet for a wide range of environmental angst. "People are so tired of hearing about oil spills and nuclear accidents and ozone—things they can't do anything about," Mielke observes. "Recycling is the way they can do their part."

Paper Glut

Alas, it isn't that simple, as Minneapolis discovered last spring [1989]. Thousands of residents eagerly turned in their glass, cans, and newspapers. But newsprint handlers were so inundated that rather than buying it for $12 a ton, some started charging $20 a ton just to haul it away. Success has threatened newspaper recycling pro-

grams all over the country. Only eight U.S. paper mills are equipped to turn old newspapers into new newsprint, and their capacity is still geared more to the scale of Boy Scout paper drives than mandatory municipal collection. In August, the nationwide glut of newsprint stood at 1 million tons. Industry officials say markets have improved since then, but they complain that too many cities launched into newspaper collection before securing purchasers. Washington, D.C., is among them: Papers picked up in its two-month-old recycling program are piling up in a big storage pit. If a buyer isn't found, all those carefully sorted newspapers may simply be hauled off to a dump or an incinerator.

Many other efforts to reduce the nation's trash volume are working at cross-purposes as well, leaving citizens who want to help wondering what to do and whom to believe. Sales of degradable disposable diapers are soaring; some communities now require degradable plastic grocery bags. Yet most experts dismiss such items as little more than marketing ploys that won't do much to reduce volume in landfills. Photodegradables decompose only in the presence of sunlight, which doesn't shine inside covered dumps. Many biodegradables rely on microorganisms to digest additives like cornstarch, but disintegration takes place very slowly in dry, oxygen-starved landfills. What's more, if degradables are mingled with recycled plastics, they can weaken the resulting products: Picture your fence posts made of recycled plastic sagging in a couple of years. The rush to degradable plastics "is a joke," says Jack Hogan, a group vice president of Spartech Corp., which nevertheless makes the material. "Our company is responding to our customers, who are forced to do this because of legislation. But you and I will be part of history when they degrade in landfills."

Polystyrene Predicament

Perhaps no consumer item better symbolizes the crisis—and the contradictions—than the polystyrene foam containers that keep McDonald's hamburgers warm and litter roadsides with such appalling frequency. McDonald's switched from paper to the plastic packaging ten years ago amid concern over vanishing forests and paper mill pollution and was a leader in eliminating ozone-harming chlorofluorocarbons (CFCs) from polystyrene production. Now facing restrictions on the foam containers in nearly 100 communities, the company is scrambling to recycle the material. Last month 100 McDonald's restaurants in New England began asking customers to toss their polystyrene into separate trash cans; fledgling recycling centers then pound it into plastic pellets that can be used in such things as Rolodex file holders, cassette boxes, and yo-yos. Someday, McDonald's envisions building whole restaurants out of recycled burger boxes. "This material has many, many uses," insists Ray Thompson, a spokesman for Amoco, which makes the boxes. "It only makes sense to enrich our waste stream with *more* polystyrene."

Polystyrene makers insist that their abandoned trays, coffee cups, and containers comprise less than 0.25 percent of the nation's trash. The biggest single component—41 percent by weight—is paper products, and their share has grown steadily, thanks in part to reams of computer phone books. Yard waste is the next biggest source by weight (18 percent before recycling), followed by metals (8.7 percent), glass (8.2 percent), food (7.9 percent), plastics (6.5 percent), and wood (3.5 percent). Toxic materials make up about 1 percent of the waste stream. They are supposed to be disposed of separately in facilities approved by the Environmental Protection Agency (EPA) and carefully monitored for leakage. But many hazardous household products—from paint to nail polish remover—slip through the EPA's guidelines.

*P*ractically nothing decomposes in a landfill. But the slow rate of degradation is actually a blessing. If more of the contents did decompose, toxic materials would pose more of a threat to groundwater.

Roughly 80 percent of all that stuff ends up in landfills. Some 6,000 remain nationwide, from unruly city dumps to state-of-the-art engineering marvels. Inside these landfills, some methane gas is produced, but not much else happens. "Practically nothing decomposes in a landfill," says University of Arizona anthropologist William Rathje, who has made a career excavating dumps from Tucson to Chicago. Rathje has found recognizable hot dogs, corncobs, and grapes buried for 25 years, and readable newspapers dating back to 1952. The slow rate of degradation is actually a blessing, he says. If more of the contents did decompose, that would hasten the rate at which toxic inks, dyes, and paints would mix with the leachate, posing more of a threat to groundwater.

Landfill operators say that newer facilities pose few environmental dangers. "These are not just holes in the ground," says Bill Plunkett, spokesman for Waste Management Inc., the nation's largest solid-waste disposal firm. "They are highly engineered excavations which have expensive leachate- and gas-collection systems." In addition to taking in 2,500 tons of trash each day, Waste Management's 397-acre Settler's Hill landfill in Geneva, Illinois, recovers enough methane to power 7,500 homes. The site will include two golf courses, a driving range, jogging and biking areas, a lake, and a picnic area. Still, NIMBY reigns supreme. Even when Waste Management offered $25 million to Chicago's Lake Calumet area for permission to expand a landfill there, residents weren't swayed. "No dumps, no deals," says state Rep. Clement Balanoff. "We have done more than our share."

Incinerators draw even more ire than landfills these days. "Resource recovery" plants were the bright hope of the 1970s energy crisis, promising to provide steam and electricity while simultaneously reducing trash volume by 90 percent. Some 155 incinerators are now in operation and 29 more are under construction. But an additional 64 have been blocked, canceled, or delayed. The problems are partly economic: Construction costs can run as high as $500 million, and the energy that incinerators produce is not cost-competitive, even though public utilities are required by law to purchase the power. But what worries residents more are the toxic air pollu-

tants, including dioxins, that some incinerators release. Even the leftover ash can be toxic and should be disposed of as hazardous waste. In Philadelphia's own version of the Islip garbage barge, a ship carrying 28 million pounds of city incinerator ash was rejected by seven countries on three continents over a 22-month period. It was ultimately dumped last year, but the ship's owners refuse to say where.

Incinerator operators say that pollution controls such as high-temperature furnaces, scrubbers, and bag houses virtually eliminate harmful emissions. They and environmentalists have dueled with scientific studies over how dangerous trash burning can be. Los Angeles's Lancer project was stopped after one test found it might cause an additional 0.118 case of cancer per million people. Detroit officials say the ash from their new $438 million incinerator is no more toxic than what remains in someone's fireplace. Yet the Detroit facility has repeatedly failed state ash tests, and in September [1989], it flunked air-pollution tests with mercury emissions four times Michigan's allowable level. Allen Hershkowitz, senior scientist at the Natural Resources Defense Council, believes that incineration *can* work safely. But he says too many U.S. operators see the environmental concerns as public relations problems rather than serious calls for upgrading practices. Meanwhile, says Hershkowitz, the fast promises of incinerator operators have "conspired to blind public officials to the opportunities that recycling offers."

Recycling Update

Recycling also holds the edge in creating new jobs, protecting the environment, and conserving natural resources. EPA has set a goal of recycling 25 percent of the nation's waste by 1992. But success will depend largely on finding markets for the recycled products. "Most people think they put out the glass, aluminum, and paper and they've recycled. In fact, all they've done is separate," says Edward Klein, who heads the agency's task force on solid waste. "Until those commodities are taken somewhere else and used again, you haven't recycled."

What follows is a status report on the potential markets for recyclable materials.

Aluminum. Turning bauxite into new aluminum is ten times more expensive than reprocessing used cans. That's one key reason more than half of all aluminum beverage cans are recycled today—42.5 billion annually. Even so, Americans still toss out enough aluminum every three months to rebuild the nation's entire airline fleet.

Glass. Reusing old glass also costs less than forging virgin materials. To date, only 10 percent of it is recycled, but markets are growing steadily. Glass bottles can live again as "glassphalt" (a combination of glass and asphalt) and, of course, as other food containers. The California firm Encore! has even disproven the old adage that new wine can't come in old vessels. It grosses $3 million a year collecting and sterilizing 65,000 cases of empties each month and selling them back to West Coast wineries.

Yard waste. Composting America's fertile mounds of leaves and grass clippings could eliminate one-fifth of the nation's waste—and as much as one-third of L.A.'s total. But aside from backyard gardeners, there hasn't been much call for mulch. Pesticides and lawn chemicals also pose toxicology problems in compost heaps. Still,

experts say that markets would grow if more municipalities followed Fairfield, Connecticut, which this fall opened a $3 million composting center to create topsoil for parks, playgrounds, and public landscaping.

Plastics. The $140-billion-a-year plastics industry is at last waking up to recycling. Currently only 1 percent of plastics is recaptured, but manufacturers are scrambling to find new uses, from plastic "lumber" to stuffing for ski jackets. Procter & Gamble is making new Spic and Span containers entirely from recycled PET (polyethylene terephthalate, the stuff beverage bottles are made of) and hopes to turn even used Luvs and Pampers into plastic trash bags and park benches. There are still limitations. The Food and Drug Administration will not permit recycled plastic to be used to serve or store food, since it cannot be decontaminated. Some products—like squeezable ketchup bottles—have up to six layers of polymers, which complicates separating. And some recycled plastic never looks new again. Tom Tomaszek, manager of Plastics Again, a polystyrene recycling center near Boston, says the pellets his plant produces come only in "army green," while manufacturers want white or clear. "The biggest problem," Tomaszek says, "is getting people away from the idea that new is better."

Paper. Industry officials like to boast that nearly 30 percent of all paper products consumed in this country are recycled—26 million tons a year, turning up in cereal boxes, toilet tissue, even bedding for farm animals. Still, that leaves more than 40 million tons clogging landfills and going up smokestacks annually. Tree-poor nations like Taiwan and Korea import some U.S. wastepaper. But matching supply, demand, and reprocessing capacity at home will take time and coordination. U.S. recycling mills can actually use more high-quality white paper, like computer printouts, than communities are collecting. For newsprint and magazines, the opposite is true. (Currently, no U.S. mills turn recycled fibers into the kind of glossy, clay-coated paper that *Newsweek* uses.) With burgeoning collection efforts, "we have the potential in this country to increase the supply [of wastepaper] overnight," says J. Rodney Edwards of the American Paper Institute. "But expanding the capacity to use it will take three to five years."

Historically, the American marketplace has been driven by *demand* for products, not *supply* of raw materials. But some lawmakers have concluded that the free market needs adjusting if recycling is to significantly dent the nation's garbage piles. Florida now taxes nonrecycled newsprint at 10 cents per ton. California and Connecticut have passed laws to require newspapers to use a rising percentage of recycled paper in coming years. Elsewhere, officials are looking for ways to force manufacturers to take more responsibility for the disposal of their products, long after they have left the factory. Some states are considering taxing products made with materials or packaging that can't be recycled. Deposits are another approach. One bill before Congress would require anyone who sells batteries to take them back and return them to an appropriate recycling facility. "We need to get the upstream side to recognize that the end of the pipeline has only so much capacity," says Lanny Hickman of the Governmental Refuse Collection and Disposal Association.

Busy, distracted citizens also need incentives to recycle. To that end, New York City plans to fine residents who don't comply with its nascent program: $25 for the first offense, $500 for four offenses in a six-month period, with landlords dunned up to $10,000 if their tenants don't cooperate. San Francisco is relying heavily on public education, reminding residents that recycling one aluminum can, for example, can save enough energy to run a TV for 3 hours. Nearby Berkeley has even tried gimmickry. Under its Eco-lotto program, city officials selected one household by random each week, rummaged through its trash, and promised to award the homeowner $250 if no recyclables had been discarded. If the lucky household wasn't careful, the cash was rolled over into the next week's prize. Sadly, the pot rose to $4,500 before the city found a winner.

Charging citizens directly for the amount of trash they throw out may prove more effective. In many communities, garbage removal still comes as part of a general-service bill, so residents never know how much it costs. Seattle changed that with "variable can rates" in the 1970s. Today residents pay $13.75 to have a single can of trash picked up four times each month, and $9 for each additional can. Faced with such fees, residents have been buying more recyclables, fewer packaged goods, and more large-size boxes. Two-thirds of Seattle households also signed up to have private haulers collect their recyclables and yard waste. Combined, those measures have cut Seattle's trash volume by 25 percent. City officials are still considering building an incinerator, but thanks to recycling, they hope it will be a smaller, cheaper one.

The Future

Seattle's program is considered a model of modern "integrated waste management"— in short, reducing and recycling as much trash as possible, then burning or burying the rest. Some local environmentalists still bitterly oppose incineration, but organizations from the Environmental Defense Fund to the Office of Technology Assessment say that the nation will need a judicious mix of all four methods to handle its trash in the future. Many experts believe that at best, the nation can recycle only about half of its garbage; recycling itself produces some residues that need to go to landfills. But those, too, can be used more efficiently. One Long Island firm, Landfill Mining Inc., has come up with a technique to "recycle" old dumps: digging through them to unearth reusable material like metal and glass. That allows the landfills to be lined and leaves room for more waste— or incinerator ash. "We can actually recycle the space," says founder Robert Flanagan.

The nation has only begun to focus its collective ingenuity on source reduction— generating less garbage in the first place. For that, citizens will need to rethink their priorities. It wasn't so long ago that Americans reused string and rubber bands, resharpened razors, threw food scraps into the stock pot, and made grease into soap. (At the turn of the century, pigs were also part of waste management in many cities; 100 of them could eat a ton of garbage a day, creating low-cost meat and fertilizer.) The disposable society has brought many conveniences. Single-serving frozen dinners in microwavable trays do facilitate mealtime. Wrapping fast-food utensils in plastic cuts down on germs. Ketchup really does come out tastier when the bottle can be squeezed. How much do those conveniences mean to us? Are they worth the price of a landfill or an incinerator next door?

Unfortunately, the trade-offs aren't so obvious. It has been far too easy, when taking out the sixth or seventh bag of trash in a week, to assume that the garbage truck will dump in someone else's backyard, or not to think about it at all. With rare exceptions during wartime, Americans have not been adept at making individual sacrifices for the common good. That mentality will have to change. Otherwise, the dumps will cover the country coast to coast, and the trucks will stop in everybody's backyard.

The Supply-Side Theory of Garbage

By Sharon Begley and Patricia King

Another day, another 3.5 pounds of garbage to generate. Sound like a challenge? It's a snap to fulfill your quota. Brush your teeth, rinse, toss the paper cup. Shave with a disposable razor. Comb your hair—oops, a couple of comb teeth snapped off; out it goes. For breakfast, a single-serving cereal box and a juice-in-a-box. There's no time to do dishes, so use a paper bowl and plastic spoon. Off to work, buying a cup of coffee and newspaper on the way. And on through the day. . . .

Every American seems to be doing his part; we're producing twice the solid waste per person as the Europeans. This dubious achievement, says Steve Romalewski of the New York Public Interest Research Group (NYPIRG), means that cutting down on trash is "a primary component in a sage and sensible solution to the garbage crisis." The less there is, the less that must be recycled, burned, or buried. Known by the unglamorous name *source reduction*, its potential is huge. Packaging accounts for about one-third of solid waste, and throwaway items such as plastic utensils account for still more: 1.6 billion disposable pens, 2 billion disposable razors, and 16 billion disposable diapers a year. "We have been a throwaway society," says Norman H. Nosenchuck, director of the New York State Division of Solid Waste. "We simply have to change our ways."

Thin Diapers, Less Packaging

Some changes are already here. In the last 20 years, the soft drink industry has cut the plastic used in 2-liter bottles by 21 percent, the aluminum in cans by 35 percent, and the glass in nonrefillable bottles by 43 percent. Each cut saved money for the manufacturer. McDonald's now pumps syrup for soft drinks directly from delivery trucks into tanks in the restaurants, rather than shipping it in disposable cardboard containers. That saves 68 million pounds of packaging a year. In the fall of 1989, Procter & Gamble announced that it is test-marketing fabric softener concentrate that the shopper purchases in small paper cartons, pours into a dispenser at home, adds water, and shakes. Those steps replace the plastic jugs of ready-made softener, reducing packaging by 75 percent. P & G's disposable Pampers diapers now come in a thin variety that does the same job with half as much material. Aveda, a cosmetics firm in Minneapolis, is designing a metal makeup bottle that the consumer can take back to the store for a refill.

Further reductions will require more radical changes. All packaging exists for a reason. Concerns about tampering led to the plastic collars around products from aspirin to yogurt. Waxed-paper inner bags keep cereals in cardboard boxes fresh. But many products owe their existence to the quest for convenience and to marketing ploys. Today we use microwavable, throw-away trays of frozen foods rather than casserole dishes, and buy microwave cake mix complete with a throwaway baking pan. Our kids tote juice boxes to the playground and play miniature Helmut Newtons with their Kodak Flings or Fuji disposable cameras. Toothpaste tubes come in boxes so stores can stack them easily; cereal boxes are bigger than needed so they make an impressive display on the store shelf. Cookies come neatly arranged in plastic trays inside paper bags. The list is as long as a supermarket aisle.

Americans won't give up their disposable lifestyle easily, but a little old-fashioned Yankee ingenuity can help. Office workers might make photocopies on both sides of the paper, halving the amount intended for the memo tray as well as, eventually, the circular file. But short of having the state set wages, there is little hope of making it cheaper to repair a radio than to buy a new one.

Reducing the Burden

What if consumers had to pay up front for their wastrel tastes? The Environmental Defense Fund, an environmental research and lobbying group, has proposed a sales or user tax based on the quantity of packaging in a product and a national sales tax on disposable items like diapers, razors, and plates. NYPIRG prefers a deposit on packages, to induce consumers to opt for minimal packaging or to at least return the containers for recycling. Rhode Island taxes fast-food packages already; the revenue is earmarked for a litter cleanup program. None of the changes will be easy. Politicians like taxes even less than they do garbage dumps. And the livelihood of some families depends on hawking disposable razors. But reducing garbage at the source will lessen the burden on incinerators, recycling programs, and landfills, to say nothing of bringing a little white-cloth gentility to the school lunchroom.

In New Jersey, and in Earnest, a Museum of Trash

By Anthony DePalma

Over the last half century, so much trash has been dumped in the Hackensack Meadowlands in New Jersey that garbage has become part of the very fabric of life here.

There is a street in the meadows called Disposal Road. Most new office buildings and warehouses sit on several feet of trash. The 150-foot-high hills that stick out of the swamps just a few miles from Manhattan are pyramids of garbage.

And now the flood of society's waste has found its way into the great dust heap called history, in the form of New Jersey's first museum devoted to trash.

Garbage specialists are confident it is also the nation's first.

Deep inside a Dump

Scheduled to open in the fall [1989], this museum of modern disposal, housed in the six-year-old environmental center in Lyndhurst, will give visitors a sense— though not of smell—of being deep inside a dump. It will explain how the trash got there and what conscientious consumers can do to cut down the ton of it that the average New Jerseyan produces each year.

"We want people to start thinking in a way they hadn't done before," said Anne E. Galli, director of environmental operations for the Hackensack Meadowlands Development Commission, which will operate the new $400,000 museum.

The garbage museum will be in the environmental center of DeKorte State Park, which is already pretty unusual for a park. One side of the center faces a mountain of garbage, and the other opens on a sea of reeds and water filled with egrets, muskrats, snapping turtles, and carp, with the Manhattan skyline behind them.

Visitors to the museum will walk through a cross section of a dump, surrounded by a sobering accumulation of household junk lining the walls, spreading over the floor, and seeming to crawl across the ceiling.

Climbing the Walls

The sheer mass of the old telephones and milk containers, the broken toys, the rusted car fenders, the stacks of newspapers, the bald tires, and the glass bottles is intended to give museum goers pause.

"They'll feel that the garbage climbing up the walls is overwhelming and at some point might fall over," said Robert Richardson, a 30-year-old Newark artist who was

11

Anne E. Galli (left), director of environmental operations for the Hackensack Meadowlands Development Commission, with Robin Anderson, project manager, at the trash museum in Lyndhurst, New Jersey. (Keith Meyers/*New York Times*)

commissioned to create the faux-dump. "That's good."

Richardson said he prowled the streets of Newark for raw material, often going out at 4:00 A.M. to avoid the stares of his neighbors in the city's Ironbound section. "Making art using garbage didn't make much sense to them," he said.

Richardson, who used to paint dinosaurs and make fish models for the American Museum of Natural History, said he was allowed to include almost anything he wanted. "They didn't want stinky garbage," he said. Commission officials also did not want a plastic mask of Richard Nixon because it might be considered a political statement, but Richardson included it anyway, covering it with a plastic bag.

Facing the Enemy

Just past the dump gallery are displays that try to put the mountain of refuse into perspective. Peepholes open on a domestic scene of a happy family at its wasteful twentieth-century habits, using aluminum foil and plastic cups once and throwing them away. The faces of the figures are mirrors that reflect the viewer's own.

A biodegradability display features piles of products headed for the dump. Viewers can lift a flap to see what will be left after 100 years. The cereal boxes and rotten tomatoes have disappeared, but the plastic forks and soda bottles remain.

The Hackensack Meadowlands Development Commission was established in

WHAT YOU CAN DO

Reduce paper wastes at work by circulating and posting memos instead of making multiple copies. Also copy documents on both sides of the paper. This will save file space, paper costs, and mailing costs, while reducing wastes.

Rodale Institute

1969 to manage garbage disposal in the 32-square-mile Meadowlands region. "The perception of most people," said Anthony Scardino, Jr., its executive director, "is that you put your garbage in a trash can, on a certain day of the week you put the can out on the curb, and, magically, it disappears."

But there is no magic. More than 100 communities once dumped their garbage in Meadowlands. Now only a dozen use the one remaining active landfill, and that

too may be closed in a few years. All other trash has to be trucked out of state.

Scardino said he hoped the new museum would make young people aware of the problem and help them come up with "solutions a lot of us haven't figured out yet."

 EARTH CARE ACTION

Pennsylvania Town Finds a Way to Get Locals to Recycle Trash

By Bill Paul

As millions of Americans still do, Meredith Campbell used to throw things away without giving it much thought. But now that she has to pay $1.50 for each 40-pound bag of trash she sets out at the curb, Campbell, like most people in the tiny town of Perkasie, 25 miles north of Philadelphia, has a new attitude.

These days she and her husband and 12-year-old daughter save newspapers, cardboard, glass bottles, and aluminum cans. The town provides white buckets for

these recyclable materials and picks them up, free of charge, as an incentive to recycle.

The Campbells have made some other changes, too. They compost food waste, leaves, and grass clippings in their backyard. They buy fewer rolls of paper towels. They avoid cosmetics and other elaborately packaged products. When they buy soft drinks or beer, it is always in aluminum cans or returnable bottles—never in nonreturnable plastic. And they recycle their junk mail, bundling it with the news-papers.

By the Bag

Paying by the bag for garbage collection cost the Campbells about $116 last year, compared with the $145 they would have paid had the town continued to charge a flat annual fee for refuse pickup. While the recycling program is mandatory, no fines have yet been collected, and the town hopes to avoid the need for them.

"You really don't have to recycle" for fear of fines as yet, says Campbell. "But it will cost you money if you don't."

What Perkasie (population 6,200) appears to have discovered is that Americans' interest in saving a buck can overcome their indifference to the garbage crisis.

Since 1978, more than 70 percent of U.S. landfills—14,000 facilities—have closed, primarily because of Americans' growing fear of landfill-contaminated drinking water. The Environmental Protection Agency (EPA) estimates that one-third of the remaining 6,000 landfills will close by 1992. At the same time, people are averse to air pollution and the toxic ash produced by incinerators. The newest of the incinerators burns waste to produce electricity, and they must comply with a plethora of regulations that, among other things, control emissions. But the plants produce toxic ash that must be disposed of. And most of the 100-odd municipal incinerators currently on the drawing board are opposed by citizen groups.

The cumulative effect of all this, according to solid-waste officials, is that much of the United States will be mired in garbage before the year 2000 unless source

You find that when you remove all the paper, glass, and cans, there isn't much left to throw away.

reduction and recycling become popular. But while 8 states will run out of landfill space within five years, and 15 states in less than ten years, federal and most state environmental officials haven't yet found a way to reduce the waste stream.

Perkasie, it seems, is ahead of them. Last year, when it began to charge a per-bag fee and require recycling, the town reduced by more than half the garbage it had to dispose of. Its sanitation truck collected trash just once a week, not twice, and on the second day picked up only recyclables.

Perkasians paid $81,182, or 30 percent less, for garbage collection. The town paid $100,805, or nearly 40 percent less, for garbage disposal. Older people living alone made out particularly well. Bob Kell, 73, cut back to just one bag every six weeks, or about $14 a year. In a poll taken at year end, nearly everyone in town said the program was a good value for the money. Meanwhile, three companies got a valuable source of raw material: Jefferson Smurfit Corp. (paper), Aluminum Co. of America (cans), and broker Todd Heller of Allentown, Pennsylvania (glass).

"We've let Adam Smith lead the way," says Perkasie borough manager Paul A. Leonard.

The town's official black plastic garbage bag has a tree printed on it that is supposed to be suggestive of environmental consciousness and comes in two sizes. The 20-pounder costs 80 cents, the 40-pounder

$1.50. Most people seem to prefer the larger size. Bags can be purchased at the borough hall or at local stores. Retailers report that with more people coming in, they are selling all sorts of other merchandise, too, not just the sanctioned garbage bags. Specially made, a 40-pound Perkasie bag is about three feet tall and twice as tough as a heavy-duty Hefty lawn bag, although it can still be torn by a sharp can or a raccoon's tooth.

Glass Distinctions

In addition to supplying free buckets and free weekly pickup of recyclable goods, Perkasie has a 24-hour recycling center behind the borough hall. There, however, people have to do their own sorting (brown glass in one drum, clear in another).

"You find that when you remove all the paper, glass, and cans, there isn't much left to throw away," says John Gerner, the editor of Perkasie's weekly newspaper, the *News-Herald*. Still, before Gerner puts out his garbage, he stomps on the bag, which seems to reduce its volume by a third. (Other Perkasians have trash compactors.) No matter how much a bag is stuffed, however, borough manager Leonard says nobody manages to get 40 pounds into it.

The savings reported by Perkasie make sense when you consider what Americans typically throw out. According to the National Solid Wastes Management Association, a Washington, D.C., trade group, 41 percent of U.S. residential and commercial garbage is paper and paperboard. Yard waste ranks second, at 17.9 percent, followed by metals (including aluminum), glass, food waste, and plastics.

Perkasie's recycling program currently doesn't include some metals, ferrous and otherwise. Plastic soda bottles and water jugs are being added this month under a deal with Wellman Inc., the nation's leading plastics-recycling company. Perkasie's steel cans also may soon be recycled and sold to USX Corp. Meanwhile, for people without big backyards, Leonard is considering buying a "tub grinder," a machine that chews up yard waste and even paper bags.

Anomalies in Perkasie

Still, things aren't quite what they seem in Perkasie. The town claims participation approaches 90 percent. But the figure looks high if you consider the volume of unsorted trash to be seen in apartment dumpsters. The problem: Landlords don't use the municipal system. Instead, they hire private haulers to avoid policing tenants, for whom garbage-collection costs are part of the rent. Consequently, much of what tenants toss is the same old mix of newspapers, cans, and other stuff that really ought to be recycled. Some landlords have encouraged tenants to take part by agreeing to hold down rents if tenants take responsibility for their own garbage disposal.

Some Perkasians have been burning garbage, and that is illegal. One anonymous local confesses to having a daily paper fire in his barbecue. A few businesses in town have had to chain and padlock their dumpsters. Old appliances get collected just once a month, and residents have to pay $5 for each one carted off.

At least so far, however, Perkasie hasn't had any "bagnappers." (In Utica, New York, a while ago, police said someone was dumping trash in the streets and making off with the official city bags in which it had been packed.)

Some problems Perkasie faces are beyond its control. Since Pennsylvania lacks enough landfills, Perkasie's garbage must be trucked to Ohio. Andrew Gansberg, a New York attorney specializing in environmental and regulatory issues, predicts that states will soon try to prohibit garbage imports. This issue, he says, may well wind up before the Supreme Court—as a matter of states' rights versus interstate commerce.

Regulating Incinerators

Also diverting legislators and regulators from reducing the waste stream is the politically painful process of trying to write comprehensive health and safety regulations for municipal incinerators. In Washington, the process is crawling as the Bush administration fights a Senate bill that would place such severe restrictions on incinerator emissions that it could cripple the incineration industry, according to Susan O'Keefe, the acting chief of the assessment team in EPA's Office of Solid Waste. (EPA supports a less restrictive House bill.)

Market forces, too, pose a threat, particularly declining prices for old newsprint and paper-industry opposition to recycling. While the American Paper Institute, the industry's trade group, says its members are committed to using recycled fiber,

U.S. paper companies are interested only in *exporting* old paper, thereby keeping the lucrative U.S. market for virgin fiber from their own forests. The price for recycled newsprint recently collapsed, reflecting a glut on export markets. James L. Burke, the executive vice president of Garden State Paper Co., a paper-recycling firm, says that "distasteful as it may be, there must be legislation" to force increased use of recycled fiber in domestic paper products.

Recycling, meanwhile, is catching on, albeit slowly. While the Bush administration would like to see 25 percent of the nation's commercial and residential garbage recycled or eliminated by 1992, firms such as Waste Management Inc., in Oak Brook, Illinois, doubt the figure will top 12 percent. Where recycling is mandatory, it is hard to enforce. Where it is voluntary, there is little incentive to change habits.

Growth in Costs and Tons

Between 1981 and 1987, Perkasie's cost per ton for garbage disposal rose 900 percent, to $58.95, as landfill space grew scarce and people put out more and more trash. (In 1960, Americans discarded 2.7 pounds of trash a day per person. Today, it is about 3.5 pounds and rising, according to the Solid Waste Management Association.) When town leaders proposed an incinerator, public opposition killed the idea. Meanwhile, Perkasie is seeing an influx of new residents who can't afford to live in Philadelphia's closest suburbs.

Despairing of getting federal or state help, borough manager Leonard decided to experiment. Although it wasn't the first community to try per-bag fees, Perkasie apparently is the first to have linked such fees with the threat of fines as a powerful incentive to recycle. (Pennsylvania has since required counties to develop long-term disposal plans.)

Perkasie may be sending a chill through consumer-product companies and plastics manufacturers because, as Leonard puts it, "We're not going to let the big guys dump on us anymore. Why should we pay to dispose of trash that companies needlessly produce?"

A company such as General Mills Inc., for instance, which makes Wheaties, among other products, feels trapped between consumer forces. General Mills's director of governmental relations, Larry Sawyer, says that as the population ages and as more people have smaller families, they demand more single-serving packages, and that means more paper and plastic that must be disposed of. And people are beginning to shun excess packaging because it contributes to the garbage glut.

Do Something

"We're not going to stay out on a limb forever," Sawyer says, and adds that General Mills has told its packaging suppliers to do something, although he isn't quite sure what.

Meanwhile, Meredith Campbell's refusal to buy soft drinks in plastic bottles is part of a trend. Plastic packaging is quickly becoming the number-one culprit. People know it isn't going to decompose in a landfill. "The public's distrust of our business threatens our industry," says Nicholas Pappas, an executive vice president of E. I. Du Pont Co.

Pappas argues that plastic's health, safety, and cost benefits can't be beat—and he has a point. Nevertheless, he says, "The market now demands not just that we produce useful products but also that we know how to dispose of them." Thus, the plastics industry has started pushing recycling. Du Pont, for example, has a deal with Illinois under which the company supplies highway pylons made of recycled plastic. In April [1989], Du Pont and Waste Management said they would jointly build the nation's biggest plastics recycling and reprocessing plant. A few weeks later, Browning-Ferris Industries Inc., a major waste hauler, and Wellman announced a second major plastics recycling pact.

But back in Perkasie, Leonard says he doesn't even want plastic garbage bags. He is considering switching to a heavy-duty paper bag.

Paper Profits for Boston Recyclers

By Cynthia Hanson

Five days a week, 52 weeks a year, Earthworm employees chase paper in order to pursue their ideal: Their job is recycling, and it's a profitable one.

Earthworm Inc. trucks rumble about the greater Boston area, collecting computer, copier, and other office scrap paper—20 tons per week—from schools and businesses. The 20-year-old company is nonprofit, but money from recycling and a consulting service go toward public education programs on recycling.

"What epitomizes Earthworm is that it's a combination of the idealism of the 1960s and the practical business sense to survive in the 1990s," says Rob Bauman, environmental policy adviser for the city of Boston. Other groups *talk* about recycling, Bauman says. But what he finds most impressive about Earthworm is that "they're doing it."

Earthworm competes with numerous Boston scrap paper dealers, but "they are strictly businesses," says Jeff Coyne, Earthworm's president. "As a nonprofit group, we don't want to duplicate the service of the private sector. We want to advocate more recycling, and our mission is to provide recycling services to those companies the private sector doesn't find profitable enough to handle."

A recycled calendar hangs on the wall behind him—a makeshift creation of old calendar pictures stuck to a recycled-paper calendar. The lanky blond kicks his work boots onto the desktop and eases back into his chair.

"The price of throwing garbage away has risen to a point that it's now made recycling look attractive financially," he says enthusiastically. Paper that costs $10 a ton to recycle may cost $80 per ton to throw away.

A Strong Demand

Unlike the glutted market for recycled newsprint, the market for high-quality paper scrap is strong. "If you look at the number-one exports out of the port of Boston and New York City, it's scrap paper and scrap metal," says Coyne.

Earthworm collects scrap from hundreds of businesses that generate amounts of wastepaper too small for other dealers to consider. Their pick-up service offers businesses a free alternative to throwing wastepaper away. A paper broker buys it from Earthworm, bales it, and ships it to paper mills, where it is made into low-grade paper products—cereal boxes and the like.

"I think, of all the wastepaper dealers

around, we have more accounts than any-body," says Coyne. "But, of all the companies, we probably collect less paper because we're doing smaller, less profitable stops."

They also do it more idealistically. "A lot of recyclers collect at night to avoid parking tickets and traffic downtown," he says, "but we always collect during the day because we want to have that contact with the people, with the employees, just to put a face on recycling and to answer any questions."

Their aim, he says, is to "try to change our society a little bit, to try to get people's attitudes to change toward saving instead of throwing away."

Earthworm was started on Earth Day in 1970 by three MIT students. It has been a for-profit business, a women-only recycling collective, and, finally, the nonprofit group it is today.

"What's great about Earthworm is that they have high environmental priorities, but they do it in a businesslike manner— which is often a struggle [for environmental groups]," says Jerry Powell, editor-in-chief of *Resource Recycling* magazine. That businesslike approach is rare, says Powell. Most nonprofit recyclers operate with at least some state or federal aid. Earthworm is self-sufficient.

After paying operating expenses (including modest salaries), Earthworm re-

WHAT YOU CAN DO

Encourage family, friends, colleagues, neighbors, and local organizations to recycle and sponsor recycling efforts. Recycling drives are an excellent source of income for local organizations. Collecting paper and aluminum cans can support everything from the local high school band to youth service organizations.

Greenhouse Crisis Foundation

cycles profits into public education projects. A consulting business helps put the business on a more even keel: Earthworm has consulted with the Boston Chamber of Commerce on ways to improve the amount of recycling in the business sector, and a similar project is under consideration with the New York City sanitation department.

Their educational services include a recycling library, a recycling "hotline" for the Boston area, and advice to Cub Scout packs, universities, and elementary school classes. They also consult and even give grants to communities interested in starting curbside recycling programs.

Product Labeling Efforts Are on the March Worldwide

By Tom Watson

On the wings of a West German blue angel and three Canadian doves, product labeling programs to promote environmental awareness are flying high.

Often referred to as eco-labeling programs, the government-sponsored efforts in Germany and Canada recognize products that are "environmentally friendly," or at least environmentally benign. Products in certain categories—recycled paper and mercury-free batteries, for example—can be licensed to carry the official program logo. In West Germany, which pioneered eco-labeling, the symbol is a blue angel. Canada, just starting what it calls the Environmental Choice program, uses a stylized maple leaf formed by the wings of three doves.

Japan recently launched a similar program. Norway and several other European nations are developing programs, and there is increasing talk of a cooperative labeling plan for all of Europe. Just within the past year, eco-labeling has rapidly gained momentum as a worldwide movement.

Germany's Blue Angel Program

Both the German and Canadian programs have the same basic format, which was de- veloped by West Germany: Under mandate from high government officials, the federal environmental agency sets up the program and helps administer it. The public can suggest product categories, including packaging. An independent commission makes the final decisions on which product categories should be included in the program. The commission includes representatives of industry and the scientific community, as well as consumer and environmental groups. An independent, nongovernmental institute or association helps set guidelines and certifies that products meet the guidelines established for that product category. Manufacturers pay fees for the rights to use the logo on their products. These fees cover most or all of the program's costs.

The only eco-labeling program with any track record is West Germany's Blue Angel program, which began in 1978. Although the program started slowly, in recent years the number of products emblazoned with the angel has increased dramatically. About 3,000 products in 60 different product categories now carry the logo, said Wolfgang Schirmer, director of the Institute for Quality Assurance and Labeling in Bonn, West Germany. The institute, known as RAL, is the independent organization that performs the testing and certification for the program.

Schirmer said the success of the Blue Angel program is demonstrated by the volume of applications for licenses. Manufacturers have embraced the program not for humanitarian reasons but because they know consumers look for the blue angel, Schirmer noted. "They are not the Red Cross," he said. "The producers are doing it to earn money."

Officials with several European companies have said they believe German consumers are far more environmentally aware than other European consumers, according to a report recently issued by Environmental Data Services Limited (EDS), an independent research group based in London.

Without the Blue Angel program, Schirmer said, consumers would have a difficult time trying to interpret the various claims and "self-created environmental labels" used by manufacturers. "There is a need for a neutral, independent, official environmental certification," he said. Schirmer foresees dozens of nations eventually having their own programs.

A number of West Germany's product categories are based on the use of recycled materials to make the product. According to EDS, these categories include building materials from recycled glass, plant pots from recycled materials, and wallpaper from recycled paper. Several packaging materials are also on the list, such as returnable glass bottles and reusable industrial packaging. Categories not directly related to recycling include asbestos-free brake linings for cars, products operated by solar energy, and low-noise lawnmowers.

EDS notes that the Blue Angel program has had a major impact on West Germany's recycled paper industry. However,

West Germany

some paper companies that produce products that could qualify for the logo have not applied for it, apparently because they do not want to call attention to their other products that are made from virgin pulp.

Until a few months ago, paper products only had to contain 51 percent or more wastepaper to qualify for the blue angel. But Schirmer said the standard has just been tightened, and paper products must now be made from 100 percent wastepaper. Also, the wastepaper must be post-commercial, not clean production trimmings.

The strengthening of this standard and those for certain other categories is apparently a response to criticisms that the Blue Angel program has not updated its guidelines often enough. The decision-making commission, which meets twice a year, is now trying to tighten standards for at least one category at every meeting, Schirmer said. These changes keep the standards in line with new technology and other developments, he added.

Another recent change in the Blue Angel program involves the wording that appears inside the logo. In the past, the wording stated: "Environment-friendly be- cause . . ." A reason was then given such as ". . . contains more than 50 percent re- cycled material." Environmentalists pres- sured program officials to remove the "en- vironment-friendly" tag, saying that no product is truly friendly to the environ- ment. This dispute actually went to West Germany's highest court, which eventually ruled in favor of program officials, Schirmer said. But by then they had al- ready given in, and from now on "environ- ment-friendly" will be replaced by other wording, such as "helps reduce waste."

Canada

Canada's Environmental Choice

This same issue surfaced in Canada almost immediately. Environment Canada, the government agency sponsoring the new la- beling program, had christened it the En- vironmentally Friendly Products program. But at the first meeting of the independent management board (it is sometimes re- ferred to as an advisory board, but it has management control over the program), members voted to change the name to the Environmental Choice program.

Board member David Cohen, an envi- ronmental law professor at the University of British Columbia in Vancouver, noted that marketing people love the term "envi- ronmentally friendly," but board members just didn't feel it was accurate. Environ- ment Canada still refers to "planet-friendly products" in its promotional brochure for the program, however.

Canada's program was announced in June 1988. The first three product cate- gories received preliminary approval ear-

ly last year [1989]. Those categories are re- refined motor oil, insulation material from recycled paper, and several products from recycled plastics, including office supplies and flowerpots. The first of these products sporting the three-doves logo were ex- pected to appear in stores late this summer [1990].

Six additional product categories have been proposed and were scheduled to undergo the standard 60-day public com- ment process this summer and early fall [1990]. Those six categories are fine paper from recycled paper, sanitary paper prod- ucts from recycled paper, products from recycled rubber, zinc-air batteries, low- pollution paints, and biodegradable cut- ting oils used in machine shops.

The Canadian Standards Association (CSA) serves as the certifying and testing organization for this program. CSA is a not-for-profit, independent testing and standards-writing association that could be considered the Canadian equivalent of Un- derwriters Laboratories in the United States, said Robin Haighton, a manager for CSA. Like RAL in West Germany, CSA per- forms its tasks for the labeling program under contract with the government.

A manufacturer must pay licensing fees for the rights to use the three-doves logo on an approved product. The annual fees are $1,500 for products with annual sales up to $500,000, $2,500 for products with annual sales of $500,000 to $1 million, and $5,000 for products with annual sales of more than $1 million. Applicants must also pay a one-time initial testing and certification fee, which will vary depending on the complexity of the evaluation criteria for that particular category.

The federal government is footing the start-up costs for the Environmental Choice program, but the licensing fees are expected to eventually cover all costs, according to program officials.

The program's management board has chosen to look at the environmental effects of products "from cradle to grave," said board chairperson Pat Delbridge. She is president of Pat Delbridge Associates Inc., an international issues-management consulting firm based in Toronto. The board tries to recognize products that have "a fairly significant net benefit" over competing products, she added.

But board members, as well as CSA and Environment Canada, have discovered that trying to gauge the environmental effects of a product during its entire lifetime can be extremely difficult and time-consuming. Attempting to compare different types of products adds another level of complexity.

For example, the board has reportedly hesitated to certify any beverage containers because it does not yet have solid evidence on whether plastic, glass, or aluminum is better overall, taking into account recyclability, energy use in production, and other factors.

In some cases, it has been impossible for the board or CSA to obtain important information about the environmental effects of a product. Board member Cohen said it has become apparent that some manufacturers don't know these answers because they never did an environmental impact analysis of their products at the research and development level.

Cohen said a major contribution of eco-labeling programs will be the development of a methodology of environmental impact analysis at the product development level. "I think industry will pick this up," he added. "I hope they do."

Board member Alasdair McKichan is president of the Toronto-based Retail Council of Canada, which represents large and small retailers throughout the country. "There has been zero resistance from retailers," McKichan said. "We think the program will be a useful guidepost to the consumer."

Some Canadian retailers, including two major supermarket chains, already have sections of "environmentally friendly" or "green" products, McKichan said. However, these may include products such as degradable plastic garbage bags, which are not expected to be approved for the Environmental Choice program because of unanswered questions about their ecological impact.

Efforts in Other Countries

Among Far East countries, the first eco-labeling program is in Japan, which announced the program last year. According to EDS, the first proposed product categories in Japan are books and magazines from recycled paper, compost-makers to use with organic wastes, several strainers

Japan

and filters for use in domestic kitchens, and personal-care aerosol products that do not contain chemical propellants.

Norway, Sweden, Holland, and France also have eco-labeling programs under development, EDS noted. Wolfgang Schirmer, of Germany's RAL, said he believes that as many as 25 countries will have programs in the near future.

McKichan envisions an international eco-labeling system. "I think that's the ideal," he said, adding that it would be very time-consuming and expensive to set up.

Although that may be a goal for the distant future, a more likely possibility is a cooperative European program. The removal of internal trade barriers in the 12-nation European Community is scheduled for 1992, and some see that as an excellent opportunity for the introduction of a European eco-label.

But Schirmer believes it would be a mistake to rush into a European program. He would prefer that countries first try to agree on environmental standards product by product.

Because every country has its own priorities, an eco-labeling program that crosses the borders of many nations may not be feasible or even desirable, Schirmer suggested. "There are so many differences from country to country," he said.

From *Resource Recycling* (Portland, Oregon), October 1989. Reprinted by permission.

EARTH CARE ACTION

"Slot" Machine
Adds Reward to Recycling

Even in an era when environmentalism is becoming trendy, consumers need a boost. So how about this: Recycle your empty soda can and get a coupon good for a full one.

Now, two seemingly incompatible pastimes are united in a reverse vending machine that aims to promote recycling by offering incentives in a way that resembles gambling on a slot machine.

Consumers who place an aluminum or tin can in the free-standing "Lucky Can" set the spinning wheels behind five display windows in motion. When the wheels stop spinning, consumers will receive a printed reward in the form of a coupon, voucher for merchandise or services, or an environmental theme message.

The machine, invented and produced by Egapro Management A. G., Zurich, has been tested in Austria, with scheduled rollouts in the United States, United Kingdom, Italy, the Netherlands, and West Germany.

When tested for eight weeks in Linz, Austria, more than 1.2 million cans (more than 1,000 cans per day per machine) were collected at service stations and supermarkets. The program is expected to roll out nationally in Austria by this summer [1990] and to increase the can recycling rate in that country from under 10 percent to more than 50 percent. Each unit can store up to 2,000 cans.

The Howard Marlboro Group, New York, has exclusive U.S. rights for the machines. Vice president Rich Wilson says four levels of sponsorship are being offered in the United States, where firms can opt for national or local participation. These are total sponsorship, cosponsorship with one other company, or sponsorship of one position on the wheel. The final option, most suitable to retail outlets, is to buy the machine and sign on sponsors yourself. (All rates are currently negotiable.) Funds raised from the recycled materials will go solely to the sponsors or be shared with the owner of the machine's location, depending on each deal's structure.

Future machines will be able to read bar codes and track which cans are being inserted in the machine. A version in production for use at golf courses will dispense golf balls.

Also in development, says Wilson, are machines that will accommodate glass, plastic, and batteries. All distribution channels are being explored, plus schools and military bases—anywhere people may drink beverages.

Although vandalism could become an issue, Wilson believes the machines will remain intact once people realize nothing of intrinsic value is inside. Coupons and vouchers aren't printed until a can is deposited.

As for the social benefit, he says, "You have mandatory legislation in some places, and people still aren't bringing back that many cans."

From *Adweek,* January 8, 1990. Reprinted by permission.

GREENHOUSE EFFECT

Engineers for Pacific Gas & Electric Co. set up solar panels at the company's solar test facility in Davis, California. (Photo courtesy of Pacific Gas & Electric Co.)

THE HEAT IS ON

By Oliver S. Owen

Consider the following events for a moment.

➤ In 1987, an enormous chunk of ice—twice the size of Rhode Island—broke away from the Antarctic ice field and splashed into the sea.

➤ Shortly after being spawned off the west coast of Africa in August 1988, Hurricane Gilbert attained wind speeds of more than 200 miles per hour. At its peak, it was the most violent hurricane ever experienced in the Western Hemisphere.

➤ During the summer of 1988, all-time heat records for many cities throughout the United States were shattered. On August 1, for example, Furnace Creek, California, lived up to its name with an oven-like temperature reading of 116°F (47°C).

At first glance, it would appear that the above events have little in common. Nevertheless, according to some scientists, these seemingly unrelated episodes may all be the direct result of a global climatic phenomenon called the greenhouse effect. The importance of this phenomenon to human society can hardly be overstated. Indeed, according to one prominent climatologist, the greenhouse effect, if uncontrolled, could wipe out civilization within 500 years.

It is ironic that we ourselves are responsible for the climatic dilemma with which we are now threatened. "Humanity is conducting an enormous, unintended, globally pervasive experiment whose ultimate consequences could be second only to a nuclear war," was the message issued by an international conference of scientists and policymakers that convened in Toronto in 1988.

The "test tube" that humanity is using in this global experiment is the atmosphere. Into this test tube we are spewing a variety of greenhouse gases, such as carbon dioxide, methane, nitrous oxide, and ozone, all of which have a warming influence on the world's climate. These gases are emitted from millions of industrial smokestacks, motor vehicles, waste dumps, and other sources.

27

Today the world's industrialized nations, such as the United States, England, West Germany, and Japan, are enjoying a quality of life unsurpassed in human history. Regrettably, however, that lifestyle is being bought at enormous environmental costs. And one of these costs is global warming caused by the greenhouse effect. The less developed nations of South America, Africa, and Asia are also contributing to the greenhouse problem, but on a much smaller scale.

Our inadvertent tampering with the global climate must be controlled. As one scientist at the Toronto meeting stressed: "This conference is screaming out to the nations of the world to put the brakes on the emissions of greenhouse gases."

The Greenhouse Effect Defined

What is the greenhouse effect? Let's give a familiar example. You know what happens if you park your car in the parking lot on a hot summer day and forget to open the windows. When you get back inside your car, it is hot as an oven. This rapid warm-up is due to a greenhouse effect: The sun's radiant energy easily passes through the car's windows, and some of this energy is then converted into heat or infrared radiation. Since this radiation cannot readily escape back through the windows, it is trapped inside, and the car warms up.

Molecules of greenhouse gases, such as carbon dioxide (CO_2), behave very much like the glass in car windows or in a greenhouse. In a sense, the greenhouse gases form a "glass window" over the earth. They trap heat that otherwise would escape from the earth's surface into outer space.

Carbon dioxide is continuously removed from the atmosphere by green plants during photosynthesis. On the other hand, carbon dioxide is gradually released back into the air when plants and animals respire, when organic matter decays, when forests, grasslands, or any organic material is burned, and when water evaporates. For thousands of years these processes were in balance, the amount of carbon dioxide removed from the atmosphere equaling the amount entering it. Since 1860, however, atmospheric levels of CO_2 have risen substantially. By itself, fossil-fuel consumption is responsible for the annual release of 5 billion metric tons of carbon into the air—roughly 1 ton for each person on earth! The rate of carbon release from such combustion has increased 53-fold since 1860. The clearing and burning of tropical rainforests to make room for cattle ranches and farms releases an additional 1.6 billion tons into the air annually.

Scientists have used several different techniques to determine the atmospheric CO_2 levels of past centuries. For example, they have analyzed air bubbles trapped in glacial ice and have examined wood from century-old trees. Since 1958, CO_2 levels have been continuously monitored with sensitive instruments atop Mauna Loa, a 13,677-foot volcanic mountain on the island of Hawaii. Another current monitoring site is located at the South Pole station of the U.S. Antarctic Program. Both sites are far removed from industrial areas where CO_2 levels would be abnormally high. The CO_2 levels recorded at Mauna Loa and the South Pole are virtually identical.

On the basis of such techniques, climatologists have determined that levels of atmospheric CO_2 rose from 275 parts per millimeter (ppm) in 1860 to 346 ppm in 1986—an increase of 26 percent. At current rates of increase, the CO_2 concentration will reach 550 ppm by the year 2050 and will hike up the global "thermostat" about 7°F (4°C).

The Greenhouse Controversy

It may appear that the greenhouse explanation of the global warming trend of the 1980s is universally accepted by the scientific community. This is not true. In fact, a number of highly respected climatologists are not convinced.

Stephen Schneider, a climatologist with the National Center for Atmospheric Research at Boulder, Colorado, states that the warm-up during the 1980s "doesn't prove the greenhouse effect." Schneider's views are similar to those of Chester Ropelewski, a weather expert with the Climate Analysis Center in Maryland. According to Ropelewski, "It's still not clear whether this is the CO_2 signal. The hard evidence isn't there." Kenneth E. F. Watt, professor of environmental science at the University of California, Davis, has ridiculed concern about the greenhouse effect as the "laugh of the century."

Despite such skepticism, however, several highly respected climatologists are certain that the current warm-up of our planet is indeed a bona fide greenhouse signal. For example, in late 1988, James E. Hansen of NASA's Goddard Institute of Space Studies testified before the Senate Energy and Natural Resources Committee that "we can state with 99 percent confidence that a cause-and-effect relationship exists between the greenhouse effect and the observed warming."

The Good News

The effects of the greenhouse phenomenon will not all be bad. Let's consider some of its possible benefits.

Some reduction in the cost of heating homes, stores, and factories. According to F. Kenneth Hare, a meteorologist at the University of Toronto, if the CO_2 level rises to 550 ppm by 2050, Canadian fuel costs for heating purposes could be slashed by 15 percent.

Warming of the subarctic grasslands. These areas, now populated largely by lemmings and caribou, may warm up sufficiently to attract not only human settlement but agricultural and industrial development as well. Certainly an average warming of 7°F (4°C) more would probably be most welcome to the few people who are now living in the frigid realms of northern Canada, Scandinavia, and the Soviet Union. Some greenhouse researchers have even suggested that millions of Americans will emigrate to Canada because they will find living and working opportunities so attractive. As a result, Canada's population, as well at its political and economic clout, could grow enormously, eventually even surpassing that of the United States.

Increased rainfall and a longer growing season. These factors, possible results of the added carbon dioxide in the atmosphere, should improve agricultural production in places such as Canada, Europe, and northeast Africa. For example, the 110-day growing season in Canada's Wheat

Belt could increase to 160 days. As Robert Steward, climate expert from Agriculture Canada, has said, "The greenhouse effect is not doom and gloom [for Canada] in any sense!"

The Bad News

Unfortunately, most of the effects of global warming would be highly detrimental. These negative effects include:

Melting of glaciers and rising seas. An increase in global average temperature of only 7° F (4°C) would result in a thermal expansion of the warmed-up sea water and a melting of glaciers such as the antarctic ice cap—and therefore a rise in ocean levels. Indeed, such melting has already begun, as evidenced by the slab of ice that broke off from the antarctic ice field in 1987 and splashed into the Ross Sea. So huge was this chunk—25 by 99 miles in area—that its loss actually reshapes Antarctica's shoreline. In fact, the Bay of Whales is now gone forever, except in mapmakers' memory!

Guy Guthrie, manager of the National Science Foundation's Polar Information Program, notes, "The size of the iceberg in human terms is staggering. If you could somehow transport it to California and melt it, it would supply all the water needs of Los Angeles for the next 675 years!"

A 3.3 foot (1 meter) rise on ocean levels by 2035 would cause the seas to move 100 feet (30 meters) further inland along our nation's shores, thus reshaping the coastline. Along the Atlantic and Gulf shores, major portions of Florida and Louisiana would be flooded. Billions of dollars of property, including homes, factories, chemical storage tanks, railroads, and highways would be inundated. It is estimated that the city of Charleston, South Carolina, alone would sustain $650 million in flood damage. Boston, New York City, Baltimore, Norfolk, Miami, Mobile, New Orleans, and Houston would likely experience similar destruction. Millions of people would be forced to relocate; human stress, anxiety, and discomfort would be severe. The salty water of the rising seas would gradually invade brackish water estuaries such as the Chesapeake Bay, with the result being a massive contamination of breeding and nursery habitats used by valuable food fish such as red snapper, bluefish, striped bass, and flounder. Moreover, salt water would seep into water-holding layers of porous rock (aquifers) and pollute the drinking water on which millions of people depend.

Hotter summers. Several of the leading weather experts in the United States are confident that the searing heat experienced in 1988 was indeed the result of the greenhouse phenomenon. Heat records were broken and rebroken in towns throughout the Middle West, as well as in New England, the Middle Atlantic states, and the far West. In Ohio, the Environmental Protection Agency (EPA) issued an ozone warning because the hot air acted like a lid over Cleveland, Columbus, and other cities and caused a serious buildup of the health-threatening gas. Health authorities throughout the United States advised the aged and those with heart or respiratory problems to remain indoors and reduce their physical activity during extreme heat waves. Despite these warnings, however, a number of heat-related deaths occurred.

More frequent and severe droughts.

Suppose that the U.S. Weather Service announced tomorrow that the nation would receive 40 percent less precipitation than normal during the next 100 years. The report would send shock waves through the country. Such a drastic decline in moisture would cause an environmental/agricultural/economic disaster of enormous dimensions. But such climatic trauma is precisely what is predicted by Walter Orr Roberts, former director of the National Center for Atmospheric Research in Boulder, Colorado, when the greenhouse effect really takes hold. In other words, droughts would be considered "normal" and could be expected year after year.

An increase in the frequency and severity of dust storms. The most destructive soil storms in U.S. history raged during the Dust Bowl era of the 1930s. Coffee-colored dust clouds more than a mile thick billowed darkly across central U.S. prairies. On May 11, 1934, one of these "black blizzards" airlifted 300 million tons of fertile soil—an amount equal to the tonnage scooped out of Central America to form the Panama Canal. The soil loss in this single storm was equivalent to the removal of 3,000 farms of 100 acres each.

Soil erosion is still a critical problem today. In March 1989, the U.S. Soil Conservation Service reported wind-erosion rates of at least 15 tons per acre per year on 4.7 million acres of cropland—an area the size of New Jersey. This is triple the rate that soil scientists consider "tolerable."

And what of the future? Unfortunately, because of the greenhouse effect, soil scientists predict that the dust storms of the future will be even worse. Greenhouse expert Roberts writes, "The Dust Bowl of the 1930s was the greatest climatic disaster in the nation's history. But it will seem like child's play compared to the Dust Bowls of the 2040s."

More frequent and severe hurricanes. The energy that powers a hurricane is derived from ocean waters that have warmed up to at least 80°F (27°C). Is it possible that 1988's Hurricane Gilbert was spawned by an ocean warm-up caused by the greenhouse effect? Several leading climatologists are of that opinion. One of them is Carl Wunsch of MIT. Wunsch believes that Hurricane Gilbert may have been only the first of an increasing number of very intense hurricanes that will be triggered by global warming. Wunsch has stated that "the climatic conditions associated with Gilbert were consistent with what you would expect to see happen under the greenhouse effect. My gut feeling is that that is what we are seeing." Several other researchers at MIT predict that an ocean warming of only 5°F (3°C) would fuel super-hurricanes that could be 50 percent more destructive than those of the past.

A growth in the number and severity of forest fires. The U.S. Forest Service called the summer of 1988 the worst fire season in 30 years. That summer's drought transformed much of the nation's timber into kindling wood, vulnerable to being set ablaze by lightning strikes. By midsummer, scores of forest fires were raging in Alaska, Idaho, California, Oregon, Utah, Wyoming, and Wisconsin. More than 3.65 million acres—an area larger than Connecticut—was reduced to char and smoking embers. At Yellowstone National Park alone, flames seared more than a million acres, forcing the evacuation of thousands of tourists who had come to watch the eruptions of Old Faithful. Fire Chief Fred

Roach, who had battled such blazes for more than 20 years, told reporters he "had never seen anything as awesome as this."

The outbreak of fires in 1988 may or may not have been an authentic greenhouse signal. However, scientists predict that as forests become hotter and drier when the greenhouse effect does arrive, the forest wildfire picture in the United States will be much like that of 1988, but on a regular basis, year after year.

Wildlife extinction. Many major temperature shifts have occurred during the 3-billion-year history of life on this planet. During the last thermal rise, thousands of years ago, a number of warm-climate species expanded their ranges northward as far as Canada. Osage oranges, for example, grew near Toronto; wild pigs flourished in Pennsylvania. Of course, many plants and animals were not able to adjust to the thermal shift and became extinct. The crucial point is that these extinctions were caused by a warm-up of 9°F (5°C) over a span of thousands of years. Today, however, greenhouse scientists predict an equivalent temperature rise in only 61 years.

If the global temperature does indeed rise 9°F (5°C) by 2050, the climate will shift poleward about 200 miles per century. Could wildlife keep pace? Many of their natural migration paths or escape routes would, of course, be blocked. For example, it would be extremely difficult for a deer to thread its way through the urban sprawl of a major city.

If the greenhouse prophets are correct, the rate of wildlife extinction will be much greater during the sudden temperature rise of the next century than it was in the much more gradual thousand-year warm-ups of the past.

What Can Be Done?

What can be done to mitigate the greenhouse effect? In 1989, the EPA advocated a number of policy options that, in aggregate, could reduce the rate of global warming by 60 percent, to about 2°F (1°C) per century. Among the recommended options are the following:

➤ Reduce CO_2 emissions from cars by switching from gasoline to cleaner-burning fuels such as methane, making more extensive use of mass transit, mandating that all new cars have a minimum fuel efficiency of 50 miles per gallon, and eventually converting to electric-powered vehicles.

➤ Impose a CO_2-user tax on all fossil fuels.

➤ Promote energy conservation by recycling paper, glass, and metals and using garbage and crop residues as sources of fuel.

➤ Greatly expand the development of solar energy so that the use of fossil fuels can be reduced.

➤ Develop large "energy plantations" of fast-growing trees. Burning this wood would not result in any net increase in atmospheric CO_2, since the CO_2 released would only equal the amount taken in by the trees when they were alive.

➤ Halt the destruction of forests in the tropics.

These measures should be pursued vigorously as soon as possible, both in the United States and around the world. Since CO_2 and other greenhouse gases have long air-lives and disperse readily through the air from one nation to another, the effective control of the greenhouse problem demands a concerted international effort. To this end, the United Nations has sponsored a series of conferences involving more than 40 countries. The first was held in Washington, D.C., in early 1989. It is hoped that these conferences will lead eventually to an International Law of the Atmosphere, with provisions for the significant reduction of greenhouse gas emissions throughout the world.

Unfortunately, there are serious obstacles to the implementation of such a law. One problem is that a number of underdeveloped nations, such as China and India, have just launched an industrialization process that they hope will lead to a better quality of life. However, such development is dependent on energy derived from fossil fuels, the consumption of which releases large amounts of CO_2. Currently, the per capita emission of CO_2 by heavily industrialized nations such as the United States and Japan is 20 to 50 times greater than that by underdeveloped nations such as China and India. Certainly it only seems right that the poor nations have the same opportunity to develop economically as did First World nations. Suppose, however, that China would like to increase per capita gross national product to just 15 percent that of the United States. To accomplish this, the nation would have to burn so much fossil fuel that the increase in CO_2 emissions would equal the total CO_2 released from all the coal currently consumed by the United States.

The world's soaring population is another major barrier to the international control of CO_2 emissions. The number of people on earth is projected to double in the next half-century. It is apparent, therefore, that without creative control strategies, emissions of CO_2 will increase dramatically even if the global quality of life is merely maintained at its present level.

Despite these formidable problems, a growing number of the world's weather experts, environmentalists, and lawmakers are convinced that humanity must come to grips with the greenhouse effect. At a recent congressional hearing, Sen. Max Baucus of Montana sounded a note of urgency: "The Environmental Protection Agency's policy options report makes a compelling argument for action *now*. The question confronting us is, Will we heed this warning?" We should indeed—and without further delay. After all, "the heat is on."

WHAT YOU CAN DO

Cars and trucks are major sources of such greenhouse gases as carbon dioxide—American automobiles spew roughly 1 billion tons of it into the air each year. To help reduce these emissions, walk, bicycle, carpool, and use public transportation whenever possible. Keep your car tuned up and consider fuel efficiency if you're buying a new car.

Jeremy Rifkin/Viewpoint

From *The Futurist*, September/October 1989. Reprinted by permission.

The Future of Power

By William J. Cook

Surely Faust would recognize the bargain that developed societies like the United States have struck. Burning vast quantities of coal, oil, and natural gas makes possible America's abundant lifestyle. Yet these fossil fuels shroud U.S. cities with persistent smog and shower sulfuric acid on forests and lakes. Worst of all, the billions of tons of carbon dioxide spewed into the atmosphere each year have the potential, many scientists fear, to make the earth's climate far less hospitable.

Unlike the legendary Faust, however, who sold his soul to the Devil for knowledge and power, we may be able to escape this predicament by changing to an entirely new, inexhaustible, virtually pollution-free energy economy based not on carbon but on hydrogen. The scheme, which would eliminate carbon from the fuel cycle, is fundamentally simple: Large arrays of photovoltaic cells, similar to those that now power pocket calculators, would make electricity from the sun's rays. The electric current would then be used to break down water molecules into their constituent elements, two atoms of hydrogen and one of oxygen. The hydrogen gas, when burned as fuel, would recombine with oxygen from the air, and its principal by-product would, once again, be water.

The combination of electricity and hydrogen could "satisfy all consumer and industrial needs," argues David Scott, a leading hydrogen proponent at the University of Victoria in British Columbia. Electricity from renewable sources like hydro and solar will continue to dominate in areas where it does today, powering lights, electric motors, and communications. But because hydrogen, unlike electricity, can be stored and transported in pipelines, it eventually could be used to power vehicles that are now fueled by refined oil. Either could provide heat for buildings. Given concerns about global warming, Scott believes there is "zero uncertainty" that a hydrogen economy will evolve over the next several decades. "What the hell else is there?" he asks pointedly.

This new energy scenario is not just a matter of dreamy speculation. In fact, sharp reductions in the cost of producing solar electricity make it a distinct possibility. In California's sunny Mojave Desert, Luz International is producing commercial solar electricity in the world's largest plants at prices only slightly above those for a new coal-burning system. Luz uses mirrors 17 feet high and 80 feet long in the shape of parabolic troughs to focus the sun's energy on stainless steel tubes filled with oil. Heated to 735°F, the oil turns water to steam that drives conventional turbine generators.

Since 1983, Luz has built eight solar

*T*he cost of solar technology is plummeting. A cell capable of generating 1 watt of electricity cost more than $100 in the early 1970s; today, it costs about $4.

power plants able to generate 275 megawatts of power for Southern California Edison, enough to meet the needs of 385,000 people. Six more plants coming on line by 1994 will raise the total to 680 megawatts, two-thirds the capacity of a typical nuclear power plant. Technical improvements and regulatory changes will allow larger plants. Within five years, says former Luz vice president Paul Savoldelli, they will be the cheapest source of electricity in sunny areas "regardless of oil prices or environmental regulations." Significantly, electric utilities are major investors in the Luz facilities.

Operating only when the sun is shining, the Luz solar plants can meet Southern California's highest electric demand, typically at midday when everyone is running air conditioners. In the future the plant might use a portion of its daytime solar energy to make hydrogen, which could be stored to fire the boilers at night.

Photovoltaic Magic

The cost of a second solar technology—silicon photovoltaic cells that convert photons of light directly to electrons—is also plummeting. In the early 1970s, a cell capable of generating 1 watt of electricity cost more than $100, making the exotic technology useful for powering space satellites but not much else. Improved manufacturing

dropped that cost to $20 a decade ago, and to about $4 today. This spring, Solarex, the largest U.S.-owned manufacturer of photovoltaic cells, will begin producing 1-foot-square modules made of ordinary window glass coated with extremely thin films of silicon. Large-scale production of these modules, expected in about three years, will result in power costing only $1 a watt, says vice president David Carlson, making the technology more widely applicable.

By the end of the decade, Carlson estimates, Solarex will be producing modules 2 feet by 4 feet that will cut the cost to 50 cents a watt. By the mid–1990s, he says, photovoltaics will be widely recognized as an economical way to make electricity and should attract a wealth of investment. And in the early part of the next century, he predicts, photovoltaic/hydrogen systems will begin to replace conventional fossil-fuel ones.

Though a solar/hydrogen economy would cause virtually no air pollution, the technology involved is not problem-free. Storing hydrogen in a vehicle is the greatest challenge. At normal temperatures, hydrogen gas is not nearly as dense as liquid gasoline and requires larger fuel tanks. Mercedes-Benz has equipped a conventional sedan with a fuel tank in which hydrogen is absorbed chemically into metal powder. When the powder is heated, hydrogen gas is released. The car's range, however, is only about 75 miles. Researchers believe that an automobile specifically designed to use hydrogen could go substantially farther between fill-ups. Although hydrogen is explosive, as the airship *Hindenburg* dramatically illustrated, it is no more dangerous than gasoline if used with care.

The location of solar collectors must also be worked out. Sunlight is not a concentrated energy source, so huge collector fields would be required in arid areas where the sun shines brightly. Robert Williams of Princeton University's Center for Energy and Environmental Studies has calculated that to replace all the oil used in the United States with hydrogen would require solar collectors covering 24,000 square miles, an area roughly one-fifth the size of New Mexico. Still, installing collectors in just 2 percent of the world's deserts could replace all of the fossil energy used worldwide.

And deserts are not the only potential sites for photovoltaic collectors. Solarex, for instance, is working with a roofing company to develop dual-purpose collectors that not only generate electricity but act as shingles to keep out the weather. As Carlson notes, "There are a lot of roofs."

Transporting hydrogen via pipeline from desert areas would be much cheaper than sending electricity long distances by wire. In populous areas, observes John Appleby, director of the Center for Electrochemical Systems and Hydrogen Research at Texas A&M University, hydrogen could be converted back to electricity in battery-like fuel cells that combine oxygen from the air with hydrogen to produce electricity. Advanced versions of fuel cells that now provide the electricity aboard space shuttles would also be ideal as lightweight power sources for electric cars like the one designed by Paul MacCready and developed by General Motors (see page 37).

The transformation to such a different energy system would be a monumental undertaking and would take years to accomplish. Given America's immense appetite for energy and enormous existing infrastructure based on fossil fuels, even a vigorously growing solar/hydrogen economy is not likely to make much of a dent until well into the twenty-first century, especially since conventional energy today is extraordinarily cheap. Adjusted for inflation, gasoline prices in 1988 were lower than at any point in history, and increases have been minimal since then. Electric rates have fallen absolutely in recent years as well; when adjusted for inflation, electricity now costs about the same as it did in 1975.

Even so, the benefits of moving to a hydrogen economy would be enormous. Oil imports, which now cost the United States $50 billion a year, could be cut drastically. The energy-short Third World could develop on the new pollution-free model, completely bypassing the problems coal and oil have created for industrial societies. Most important, solar/hydrogen offers industrial civilization a way to fuel a comfortable lifestyle without destroying the earth in the process.

Paul MacCready: Driving Ahead

By William J. Cook

Paul MacCready is an energy minimalist, a man driven by the ideal of propelling objects with the least imaginable effort. That quest has made the 64-year-old engineer a beacon to watch as the nation seeks to develop vehicles that consume little polluting energy.

As a youth, MacCready set records flying model airplanes, and as a young man he was three times the national champion—once the world-title holder—in sailplanes, those exotic aircraft powered only by rising air currents. He became famous in 1977, when his fragile Gossamer Condor became the first human-powered plane to win the prestigious Kremer Prize by taking off using only the muscle power of the pilot, clearing a 10-foot hurdle, and executing a figure-eight turn between two pylons half a mile apart.

Future Commuting

The hobbyist's early achievements set him off on a scientific course that could ultimately change everyone's commuting style. In January [1990], General Motors unveiled a pollution-free electric car that performs dramatically better than any before. It is the progeny of MacCready's small Southern California skunk works, Aero Vironment, Inc.

The Gossamer Condor, now displayed in the National Air and Space Museum, was only one of the elegant aircraft MacCready and his team have used to explore aerodynamic limits through low-energy designs. In 1979, his Gossamer Albatross became the first human-powered craft to fly the English Channel. Two years later, MacCready linked up with Du Pont to demonstrate that with lightweight materials and innovative engineering, solar energy alone could be used to power a plane. The Solar Challenger flew from Paris to England, deriving electricity to drive its propeller from 16,128 photovoltaic cells mounted on its wings and tail.

MacCready's exploits first attracted General Motors in 1987, when it decided to enter a race across Australia for solar-powered cars. GM put its Hughes Aircraft division in charge, and a Hughes official, one of MacCready's former classmates at the California Institute of Technology, enlisted MacCready's assistance. MacCready and Aero Vironment quickly designed and built the teardrop-shaped GM SunRaycer—a featherweight speedster, covered with solar cells, that easily outdistanced the competition. GM officials were so impressed that they bought 15 percent of Aero Vironment.

MacCready also convinced GM that

Paul MacCready and
GM SunRaycer at
proving grounds.

with careful attention to every system, from aerodynamics to tire rolling resistance, it would be possible to make a practical electric car. With Aero Vironment as primary contractor, GM built the remarkable Impact, an experimental, battery-powered two-seater with the pickup of a sports car and the driving range to satisfy most urban commuters.

Will GM Be First?

MacCready believes that with improved batteries or even fuel cells that produce electricity by combining oxygen from the air with hydrogen (a prospect he finds "especially charming"), the odds are at least 50 percent that within 15 years "electric propulsion will dominate the new-car field." He's even more certain that something like Impact will be available to consumers within the next few years. "The question," he says, "will be whether it has a General Motors nameplate on it."

The giant corporation has been wrestling with that issue all spring. This month [April], outgoing GM chairman Roger Smith took Impact to Washington to show it off. If GM elects to market a practical electric car along the lines of Impact, MacCready's relentless quest to do more with less could mean that future urban residents will breathe easier, yet still make it to work.

From *U.S News and World Report*, April 23, 1990. Copyright © 1990 *U.S. News and World Report*. Reprinted by permission.

A Tinkerer's Dream: Fueling Up the Car at the Burger King

By Amal Kumar Naj

To the white-haired septuagenarian at the wheel of the 1979 Volkswagen Rabbit, this is the ultimate joyride.

After rounding a bend, the driver guns the engine. The speedometer inches toward 75. The foliage is resplendent. And the autumn air is as crisp as . . . a french fry?

"Smell it," says the driver, sniffing the fumes from the tailpipe with great satisfaction. "It's as good as gasoline. Right now, you're riding in the world's only vegetable-oil car."

We are riding with Louis Wichinsky, environmentalist and tinkerer. He aims to take vegetable oil from the deep-fat fryer to the fuel tank, because he believes vegetable oil to be less polluting than highly touted alternative fuels such as methanol and ethanol. In his dreams, he sees millions of cars converted to run on vegetable oil.

Filtering Out the Fries

Veggie cars are reliable and get good mileage. Wichinsky's Rabbit typically gets 54 miles to the gallon, roughly the same as it gets on diesel fuel. He has gone as far west as Las Vegas cruising on vegetable oil. Best of all, in Wichinsky's view, vegetable-oil cars give off no toxic emissions.

Vegetable oil isn't exactly cheap. At $2.60 a gallon wholesale, it costs more than twice as much as gasoline. But the frugal mechanic has come up with a way to power his car with *used* oil, courtesy of his neighborhood diner and the local Burger King.

Fueling up is a bit more problematic—he has to heat the oil and filter out the stray fries. Recycled oil, though messy, has several advantages. It's free. Filling stations are as accessible as the nearest fast-food joint. And the national diet being what it is, supplies are usually plentiful.

"The only thing you have to adjust to is the smell of McDonald's on your tail," says Richard Sapienza, a chemist at Brookhaven National Laboratory in Upton, New York, who has studied the combustion of vegetable oil. Sapienza says that vegetable oil, when burned, doesn't give off the sulfur dioxide or hydrocarbons that gasoline does.

The outing with Wichinsky starts with a pit stop at the Miss Monticello Diner,

near his home in Hurleyville, New York. He is a regular, though not at the lunch counter. Typically, he heads for the drums of discarded oil behind the diner, but he is out of luck today. The drums are nearly empty.

Down the road is the Burger King. But when Wichinsky lifts the lid of the restaurant's dank dumpster, all he finds is an empty waste-oil container. A company that recycles oil for industrial uses has beaten him here.

Fortunately, Oscar Patton, manager of the Miss Monticello, comes to the rescue with 2 quarts of salad oil. "Louis, you're back again," Patton says, wrinkling his nose. "I can smell that french fry coming down the road."

Wichinsky is used to being seen as an oddball. A lifelong tinkerer, he has in-

vented a bagel-making machine (3,600 bagels an hour) and an egg separator. Neither was a big commercial success, but the inventor says he isn't in this business for fame, glory, or financial gain anyway.

Protecting the environment is a top priority. One of his current projects is a receptacle for plastic waste that would dispense cash prizes. Basically, he has attached a Bally's slot machine to a dumpster. "You need some incentive" for people to recycle plastic, he says.

Wichinsky first went vegetable during the oil crisis of 1973. Back then, he recalled that Rudolf Diesel, the late German inventor, had tried vegetable oil while running various fuels through his engine in the early 1900s. But the oil tended to clog the fuel injectors, bringing the engine to a sputtering stop. Wichinsky says he has solved that problem by injecting into the engine doses of a "super cleansing agent," a mixture of methanol and water.

A Connoisseur of Oil

His Rabbit is the fourth car he has modified to date. A switch under the steering wheel allows the driver to shift from vegetable oil to diesel and back to oil. The car doesn't flinch. But Wichinsky isn't satisfied. "I don't like this batch of oil," he says. "The idle is a little too erratic, not as smooth as peanut oil."

"The best is rapeseed oil," he muses in the tones of a connoisseur. Rapeseed is used to make lubricants, specialty plastics, and an edible oil.

Wichinsky doesn't expect to see the nation's auto dealers plugging veggie cars anytime soon. Before mass-marketing his device, he wants the approval of the federal government. He recently submitted his design to four government agencies, but he hasn't heard back yet. The U.S. Department of Energy's Office of Energy-Related Inventions says that his innovation "is under technical evaluation."

Vegetable oil faces tough competition from other alternative fuels, most notably methanol, which is made from coal or natural gas and burns cleaner than gas. General Motors Corp., Ford Motor Co., and Chrysler Corp. are manufacturing cars that can run on methanol or gasoline. Tougher clean-air laws enacted by several states, including California, will take effect over the next few years.

Other auto companies and manufacturers of alternative fuels are pushing even more exotic fuels, such as liquid propane, hydrogen, and butanol, which can be made from molasses or corn. Several makers of photovoltaic cells continue to promote solar-powered autos. Wichinsky believes that vegetable oil is more practical than these fuels. Hydrogen, for instance, is highly explosive.

"Fill'er Up!"

It's time to refuel again. Wichinsky pulls up in front of Finkelstein & Schwartz, a food wholesaler outside Monticello. He strides past stacks of industrial-size cans of Idaho Mashed Potato, Heinz ketchup, All Trumps Enriched Bromated Flour, and Glorietta Apricot Nectar until he gets to the stockpile of Admiration Pure Vegetable Oil. A 5-gallon container is $13.

The high price of vegetable oil doesn't bother Wichinsky, though. Prices would drop if certain states—say, New Mexico, Arizona, and Texas—were ordered to grow nothing but rapeseed, he says. Growing oil beats importing it, he figures.

Cruising on Admiration Pure, Wichinsky breezes down a dirt road and pulls up in front of an old cement and stone barn owned by his friend, Otto Tolski. Here on his 50-acre farm, Tolski is growing a test patch of rapeseed for Wichinsky.

The tiny yellow flowers on 3-foot-high plants glow in the sunlight. Some plants are already sprouting thin shoots full of tiny dark brown seeds. "I would like to take these seeds and toss them into every empty field," Wichinsky says.

To Halt Climate Change, Scientists Try Trees

By William K. Stevens

Scientists, foresters, environmentalists, and government officials are seriously exploring the feasibility of an ambitious long-term enterprise: planting enough trees around the world to ease the threat of global warming.

The goal is unlikely to be realized dramatically or quickly, if ever, but rather in increments, tree by tree, plot by plot, field by field. Nevertheless, fundamental facts of nature have convinced a number of experts that widespread planting of trees, along with conservation of existing forests, is one of the surest, easiest, and least expensive ways to begin to halt or even reverse the buildup of carbon dioxide in the air.

Carbon dioxide is the gas chiefly responsible for the greenhouse effect, in which heat from the sun is trapped within the earth's atmosphere instead of radiating back into space. Dead trees release carbon dioxide into the air. There it joins with even more carbon dioxide produced by the burning of fossil fuels like coal and oil. The combination, many scientists believe, is making the greenhouse effect more intense, future global warming inevitable, and major climatic disruptions more likely.

But growing trees absorb carbon dioxide, storing the carbon part of the gas and releasing the rest as oxygen. Foresters and environmentalists therefore see large-scale tree planting not only as one way to head off global warming but also, if carried out on a crash basis, as a possible emergency solution if warming should seem to get out of hand.

At the moment, the foresters face an uphill battle to overcome the widespread destruction of tropical forests taking place in Brazil, Indonesia, and other developing countries. Experts are nevertheless working hard to find effective ways to stimulate the growth of new forests in every part of the world.

The activity is taking place on a variety of fronts. In one of the first concrete actions, the American Forestry Association, a citizens' conservation organization, has undertaken a national campaign aimed at planting 100 million new trees in American cities and towns by 1992. Although that is just a tiny fraction of all the trees in the country, proponents of the effort see it as an important beginning.

Foresters are pressing experiments in farming dense stands of fast-growing trees that suck up carbon dioxide at the maximum rate. Environmentalists are advancing an "offset" strategy, in which industrial companies would pledge to plant enough trees to absorb the amount of carbon dioxide produced by new plants that burn fossil fuels. One such arrangement, widely viewed as a model, has already been undertaken by a Connecticut company.

Over the last 18 months or so, economists and ecologists have stepped up their studies of which approaches might work and which will not, what is practical and what is fanciful. Bills now before Congress seek to promote reforestation both domestically and abroad, and the Environmental Protection Agency (EPA) has undertaken an extensive study of how best to go about the job.

A Common Atmospheric Pool

The task is not as simple and straightforward as it may seem. Most possible courses of action are fraught with questions, difficulties, and uncertainties. The expanding research efforts are trying to answer some questions, but many of the difficulties may prove intractable. In the Third World, for instance, economic and population pressures force millions of people to cut forests for fuel and fodder. Large-scale development also eats up much forest land, both in the Third World and in industrialized countries. Taxing policies in some countries provide perverse incentives to clear forests.

In Brazil, one of the countries where destruction of the tropical forest is greatest, tax rates on cleared land are lower than those on forested land, said Kenneth Androsko, the chief forest analyst for the environmental agency's climate-change group. "It costs you money to keep a standing forest," he said. Dealing with political factors like this, it is widely acknowledged, is going to be difficult.

Forests in the earth's temperate regions take in as much carbon as they release, the EPA says. In the tropics, because more forests there are destroyed than are replanted, much more carbon dioxide is released into the atmosphere than is absorbed. But both regions could contribute to the solution by planting more trees, since all trees draw carbon dioxide from a common, worldwide atmospheric pool.

The task of reforestation is also made difficult because of its sheer scale. Scientists who have studied the problem calculate roughly that to absorb the carbon dioxide released into the atmosphere by human activity, it would be necessary to plant enough trees to cover an area half the size of the United States, or more. This area, according to another estimate, is equal to about one-third of the world's combined croplands. Each year, the EPA says, an area the size of Tennessee is cleared in forests in the tropics.

Dr. Roger Sedjo, a forestry expert at Resources for the Future, an independent research organization in Washington, has calculated that it would require $186 billion to $372 billion, depending on land costs, to establish enough new forests to absorb the 2 to 3 billion tons of carbon poured into the atmosphere each year. "We spend that kind of money on national defense every year," Dr. Sedjo said, while forestation would be a one-time cost.

The size of the job appears daunting, acknowledges Dr. Daniel J. Dudek, a senior economist at the Environmental Defense Fund, a research and advocacy group, who has studied the matter. The same is true of the overall question of global warming.

"When we consider solutions to the problem, frequently people back away," he said. "It's just too awesome, too large, and it will involve too many painful sacrifices."

But in fact, he said, there are ways to break the problem down into many smaller, "sane steps" that can be taken to "start down the long road of developing alternatives and solutions." Dr. Dudek advocates the planting of trees to offset new fossil-fuel plants as one of several steps involving trees that are "so simple and so easy" that they would provide "a kind of acid test about how serious we are about managing the greenhouse problem."

"No Magic Cure"

The model effort is widely considered to be the one announced last fall [1988] by an electric utility in Connecticut, AES Thames. The company, a subsidiary of Applied Energy Services of Arlington, Virginia, is helping to pay for the planting of 52 million trees on plantations and small farmers' plots in Guatemala to offset the carbon dioxide emitted by a new generating plant at Uncasville, Connecticut.

Virtually no one who has seriously examined the question believes that planting trees would be a total solution to the greenhouse problem. It is generally seen as one strategy among several.

"There's no magic cure," said R. Neil Sampson, the executive vice president of the American Forestry Association. A number of measures will be necessary "to start the trend swinging the other way," he said. These especially include weaning the world away from fossil fuels through better energy efficiency and shifting to alternative energy sources such as solar power.

But the world essentially runs on fossil fuels, and the weaning is likely to be painful. Some experts say that if the pain proves too much to bear, and the world finds itself some years hence on the brink of climatic disaster, mass reforestation might be a relatively easy emergency measure, a stopgap that could buy two or three decades of time. The trees would begin absorbing carbon dioxide immediately and would go on doing so for 20 to 30 years or more.

Proposals for Making a Dent

"If we were to decide we had to do something quick, then a tree-planting scheme might play a significant role," said Dr. Gregg Marland, an environmental scientist for the Department of Energy at the Oak Ridge National Laboratory in Tennessee. Dr. Marland was one of the first to study reforestation as a possible solution to the greenhouse problem.

Although most people are focusing for now on relatively modest beginnings, there is no lack of suggestions. The EPA, in an investigation requested by Congress, is looking into a number of foresting measures that it believes are feasible and that, when combined, could make a significant dent in the problem.

➤ Planting up to 400 million trees in urban areas of the United States. Planted strategically around buildings, they not only would soak up carbon dioxide but would also provide shade to help reduce energy use in hot weather. The American Forestry Association's drive concentrates on this option.

➤ Offering American farmers incentives to plant more trees on erodable lands that have been set aside and kept out of cultivation in the Conservation Reserve Program.

➤ Managing existing forests more effectively to make them more dense. The EPA says the total mass of American forests could be increased 60 percent in this way.

➤ Reforesting 20 percent of the United States' highway corridors.

➤ Substituting traditional "agro-forestry" for the slash-and-burn agriculture common in the tropics. In agro-forestry, which has operated for hundreds of years in some countries, crops and trees are planted together. Tree branches and leaves are cut for fertilizing mulch. By contrast, slash-and-burn agriculture uses ashes from burned forests as fertilizer.

➤ Undertaking large-scale tree-planting programs on degraded lands in the tropics, some of them denuded and abandoned after slash-and-burn.

➤ Establishing "plantations" of fast-growing trees, especially in Third World villages. The trees would be harvested on a rotating basis in the hope that villagers would not then chop down wild trees.

Trees as Crops

In the United States, the Department of Energy's laboratory at Oak Ridge is experimenting extensively with what it calls "short rotation" plantations of fast-growing trees that are ready for harvest in ten years or less. The idea is to apply agricultural principles to trees, rotating and harvesting them like crops.

Fast-growing plantations are considered particularly attractive by some experts because young, growing trees absorb carbon dioxide at the fastest rate. Mature trees absorb much less, and some scientists believe that it would be best to leave them alone, since they store large amounts of carbon.

Young growth, on the other hand, could be harvested on a rotating basis and the wood substituted for coal and oil in many applications. The theory is that the trees still growing would absorb the carbon dioxide given off by the wood that is burned. Because no fossil fuels are involved, the result would be no net increase in atmospheric carbon dioxide.

Another step being considered is putting as much of the harvested wood as possible into long-lived products such as houses, where the carbon would be sequestered for years. The forestry association has commissioned studies to determine the effectiveness of this measure.

A citizen of the deep, this sweetlips coral fish prefers the warmth of the tropics. The unusual name comes from the habit these fish have of approaching each other with a "kiss." (Terry Domico/West Stock)

MURKY WATERS

By Maria Goodavage

Throughout history humankind has regarded the world's oceans with a reverence usually reserved for the gods. Unfathomable and endlessly bountiful, they have been the most valued resource on Earth.

But now the oceans we have loved and exploited seem smaller and less powerful. Perseverance and new technology are allowing us to see them as never before. Recent discoveries have been more fantastic than marine biologists—or even science-fiction writers—could have dreamed.

At the same time, disaster threatens them.

More than 250 dolphins with eerie smiles plastered on their ulcerated mouths washed ashore along the Atlantic Coast during summer of 1987; coastal sewage laced with industrial toxins is a prime suspect in their deaths. The bodies of aquatic birds with beaks like corkscrews, no eyes, stubs for wings, and high concentrations of selenium are found on segments of California's coastline. In Alaska's Prince William Sound, an entire ecosystem was ravaged, perhaps irreparably, by the calamitous *Exxon Valdez* oil spill earlier this year [March 1989]—the worst in U.S. history. All over the globe, nightmarish aberrations are prompting us to ask some fundamental questions: Just how much abuse can the oceans take? Is there still hope for a more pristine future, or is it already too late to reverse the damage we've done?

Scientists note that our affection for the sea could paradoxically be one of the major reasons for its present distress: More than 50 percent of the U.S. population lives within 50 miles of a coastline. We generate 150 million tons of solid waste each year, much of it finding a final resting place in the oceans, and much of it a menace to life.

Flows of hypodermic needles, sutures, and catheter bags onto New York and New Jersey beaches stunned us in 1988. The ugly refuse created a short-term crisis, but it proved to be a powerful warning against using the ocean as a universal waste tank. People who never showed concern about marine pollution suddenly realized that the ocean is not a giant disposal.

47

It's fitting, in an ironic way, that the realization of our neglect coincides with some of our greatest oceanic discoveries.

"We've been killing ocean life and trying to find it at the same time," says Kathryn O'Hara, a marine biologist with the Center for Marine Conservation in Washington, D.C. "On one hand, we're destroying our world; on the other, we're just starting to understand it."

Life without Sun

Scientists from Woods Hole Oceanographic Institution expected a desolate black expanse as they cruised just above the Pacific Ocean floor in their deep-sea submersible, *Alvin*, in 1977. After all, they were 8,000 feet below the surface, far beyond the reaches of even a faint glimmer of life-giving sunlight.

They were amazed to discover a thriving, writhing community of some of the strangest, most colorful creatures ever found in one place. That excursion and subsequent dives brought researchers face to face with dozens of life varieties never before seen by humans. Flowering yellow dandelion-like animals; giant reddish-pink tube worms measuring up to 8 feet long; odd variations of old favorites like clams, mussels, snails, and crabs—all were part of the bizarre panaroma.

Life without the sun? The question baffled scientists. But the answer surprised them even more: The exotic animals flourish around vents of water heated by Earth's core. The base of this deep-sea food chain comprises bacteria that thrive on reduced sulfur compounds spewed forth by the vents.

These oases of life, which have now been found around many of the world's hydrothermal vents—and cold seeps—need no sun, contrary to the long-held belief that the main source of energy for life is sunlight. It looks as if several of these "new" species haven't changed in millions of years.

The vents also help explain the ocean's chemical composition, including its salt content. Water circulating across the sea floor seeps down through the rocks, depositing and picking up chemicals. It is heated, then spewed out at temperatures exceeding 350°C (662°F) through chimney-like vents. The vent system is so large that scientists estimate the ocean circulates all its water through the vents—dissolving and transporting minerals along the way—every 8 million years. That's a relatively short passage of time, geologically speaking.

An equally astonishing development in our recent oceanic enlightenment has been rock-hard evidence of plate tectonics, or continental drift. By drilling into the deep-sea crust and bringing up rock samples, geologists have been able to show how Earth's crust has been ripping apart for eons.

Scientists say it's no longer speculation that North America used to be part of Europe and Africa. "The development of the plate tectonics theory has provided a revolutionary new way for geologists to understand and explain the origins of geological features," says Richard Rosenblatt, Ph.D., professor of marine biology at Scripps Institution of Oceanography in La Jolla, California.

These and other discoveries would have been impossible without technological advances like deep-sea submersibles with nimble grasping arms or ships with powerful positioning and drilling systems. But the hero of high-tech ocean exploration never gets wet: the satellite.

Satellites can reveal data such as water temperatures and the highs and lows of sea level at given locations, gathering information in one sweep around Earth that would take months and countless man-hours to collect from ships.

Researchers are even using satellites to detect blights caused by man-made pollution—an increasingly frequent task in the past few years. These omniscient eyes float unseen in the heavens, looking upon devastation and reporting our sloth.

Destroying the Deep

How could the world's vast oceans, measuring 328 million cubic miles, ever be anything but impervious to our actions? Even marine biologist and writer Rachel Carson, who mobilized environmentalists with her book *Silent Spring* in 1962, couldn't predict the effect humans would have on the expansive seas. "[Man] cannot control or change the ocean as . . . he has subdued and plundered the continents," she once wrote.

To think we could change the future of something as great as an ocean was to liken

Until recently we were able to feed the oceans toxic wastes and sewage and notice few, if any, ill effects. Why? Slow reaction time. The oceans, because of their bulk, move lethargically.

ourselves to gods. Impossible, the experts said. But here we stand, approaching a new century, clenching the trident that once belonged to King Neptune.

How did we acquire this new domain? The oceans fell under our influence because of the same code of ethnics applied in many china shops: "You break it, it's yours."

The oceanic ecosystem is not as hardy as we thought. Until recently we were able to feed the oceans toxic wastes and sewage and notice few, if any, ill effects. Why? Slow reaction time. The oceans, because of their bulk, move lethargically.

"By the time a problem becomes patently obvious it may be almost too late to remedy it," says Alfred Ebeling, Ph.D., an ichthyologist with the University of California, Santa Barbara. "We have some crisis brewing right before our eyes."

Polluted Harvests

Jonathan Swift once wrote that "he was a bold man that first eat [sic] an oyster." Swift was referring to the courage it must have taken to be the first to consume the oozy body of the naked mollusk—but in our day his message carries an entirely different meaning.

Oysters and other shellfish are filter feeders, extracting nutrients—and harmful bacteria and viruses—by pumping seawater through their gills and digestive system. Pathogens are stored in their tissue and can pose a serious threat to the health of those who eat them. Typhoid, infectious hepatitis, and acute gastroenteritis are three maladies humans risk when eating shellfish harvested near grounds where cities dump sewage.

In 1977 Congress prohibited the Environmental Protection Agency from issuing any permits for dumping sewage sludge after December 31, 1981. New York City, however, went to court and, after a favorable court decision, continued dumping sewage sludge. And during heavy downpours all over our coasts, raw sewage seeps directly into the water because of overflow or equipment failure. "It's scary as hell for business and for people who value their health," says Clifford Hillman, who owns a shrimp and oyster company on Galveston Bay, Texas. Texas has been closing its shellfish beds about a dozen times a year since 1985.

Sewage dumping is a hotly debated issue. Since sludge typically contains traces of industrial pollutants like zinc, chromium, lead, and polychlorinated biphenyls (PCBs), some scientists are working to establish a link between sewage and scourges like the mass deaths of dolphins on the East Coast. Other scientists call such connections absurd, or at least premature.

Contamination from toxic wastes is more clear-cut. Around the world's seas, toxic pollution is taking a costly and disturbing toll.

➤ Some 50 beluga whales, already on the federal endangered species list, have washed up on the shores of the St. Lawrence River in the last four years. The whales died of bladder cancer and AIDS-type symptoms. Researchers found high levels of dangerous organic residues in their tissue.

➤ A mysterious epidemic killed more than 7,000 seals in the heavily polluted North Sea in 1988. Biologists say industrial poisons there may have weakened the animals' immune systems.

➤ Most states have issued advisories limiting the consumption of fish caught in certain coastal and recreational waters.

The list continues. As toxins travel up the food chain, the contaminants become more concentrated. We're only now realizing that the top of the food chain is not always the safest place to be. But it's not easy at the bottom, either. For phytoplankton—the free-floating microscopic plants that form the basis of our food chain and biosphere—survival is harder than ever.

Plastic Perils

Rarely, though, do we have cause to feast on the harvest of our neglect. More often the foul bill of goods we've passed along to ocean dwellers is cause for appalled disbelief. Whales and turtles eat floating, shimmering plastic bags and balloons, mistaking them for living morsels. They gradually die of starvation as the plastic blocks their digestive system. Plastic fishing nets and packing straps drifting through the Bering Sea kill about 40,000 fur seals each year off the Pribilof Islands. The seals play with the plastic, get tangled, and drown, starve, or die from the deep gashes created when it tightens around their bodies.

Albatross and pelican skeletons hang like ghastly ornaments on trees all over the world, forever entangled on cast-off fishing line. It's nature's reminder that we don't have to shoot an albatross with a crossbow to have it hung around our necks to shame us.

Plastic pollution is taking its toll on every form of marine life. In 1975 it was reported that some 26,000 tons of plastic packaging and 150,000 tons of fishing gear were dumped or lost at sea every year by commercial fishing vessels. We litter our beaches with huge amounts of plastics; Los Angeles County beachgoers leave 75 tons of trash—much of it plastic—each summer week. Add to that 690,000 plastic containers jettisoned daily by merchant vessels, and we have a weighty affliction. "Plastic litter could be a major killer of large marine life in the world," says Howard Levenson, senior analyst at the Office of Technology Assessment in Washington, D.C. "The bad news is that it's not going to go away tomorrow or in 100 years unless plastics are made degradable."

While scientists are sure of the long life of plastics, they debate the life span of oil. Some think the ocean has a greater capacity to break down hydrocarbons than once believed. Others say it lingers longer, citing examples like the 1978 wreck of the *Amoco Cadiz* tanker. Oil from that spill is still interfering with fish reproduction around the coast of Brittany.

But no matter how long oil remains a menace to the environment, its immediate effect is ugly and lethal. The wreck of the *Exxon Valdez* tanker in March 1989 poured 10 million gallons of oil into Alaskan waters, coating millions of marine creatures with thick, tarry shrouds. "It's the latest tragedy, and the worst, but unless government and industry start acting now, it's far from the last," said Clifton E. Curtis, executive director of the Oceanic Society.

Fighting the Damage

Schoolchildren in California, Nevada, and Texas are creating oil spills of their own. They're pouring motor oil into basins of water and testing its effect on bird feathers and seal fur. Then they disperse the oil with wind (a fan) and try to clean it up with vacuum suction, cleansing agents, and ocean skimmers (straws, detergent, and spoons). And though they may fail to eradicate their classroom oil spills, when they travel out to real trashed beaches to clear them of garbage, they soundly succeed.

Education programs like this one run by Project OCEAN of the Oceanic Society's San Francisco Bay chapter will help to determine the long-term health of the oceans. By educating upcoming generations in how to handle the oceans with care, it's possible to hope for the future well-being of our seas.

But first we have to get through the next decade or two. "If we don't act now, it will be just that much more difficult to act later," warns UCSB's Ebeling. The task at hand is perhaps the greatest ever shared by the nations of the world.

For inspiration, we can look to the rapidly improving Mediterranean Sea, which long served as a cesspool for the 18 nations bordering it. Ten years ago, one-third of the Mediterranean's beaches were unsafe for bathers. Now, because of a multinational cleanup agreement, only one-fifth are so badly polluted—and the numbers are dropping steadily.

The United States is also beginning to listen to warnings from the deep. In 1987 we became the twenty-ninth country to ratify an amendment to the 1973 MARPOOL (Marine pollution) convention that forbids ships and boats to dispose of plastics in oceans around the world. Last year [1988] Congress passed and the President signed a law outlawing all ocean dumping by 1992. The effort to protect the marine environment is not new. Legislators have enacted dozens of laws in the past three decades, since environmentalists like Jacques Cousteau raised our consciousness about the oceans. One of the most successful of these is the Marine Sanctuaries Act passed in 1972. Sparked into action by heavy toxic pollution and, at that time, the biggest United States oil spill ever, Congress moved to provide sanctuaries where natural and historical resources could remain safe. There are seven sanctuaries in American waters today, and their healthy residents are prospering in nearly unblemished habitats.

S omeday, maybe even the most polluted shores will look like today's ocean sanctuaries, or be as untarnished as the deep-sea vent regions.

It's going to take time, patience, and money to determine what we must do to regain our healthy oceans. "We can't wait for marine animals to die or disappear before we ban a harmful pollutant," says Neylan Vedros, Ph.D., professor of medical microbiology at the University of Califor-

---

WHAT YOU CAN DO

When possible, use biodegradable, nonphosphate detergents and other cleaning agents. Phosphates and other chemicals wreak havoc on aquatic life.

Center for Marine Conservation

---

nia, Berkeley. "No price can be too high for such preventive medicine."

Research into how pollutants affect the environment will have to be extensive, scientists say. And since we can't realistically banish all ocean dumping, researchers are beginning to focus attention on how to dispose of wastes more safely.

Someday, maybe even the most polluted shores will look like today's ocean sanctuaries or be as untarnished as the deep-sea vent regions. We'll be able to swim at any beach, then dine on our favorite seafood without the fear of becoming ill. The seas will again be Earth's most venerable resource.

Only then will we have fulfilled our Neptunian duties.

From *Modern Maturity,* August/September 1989. Reprinted by permission.

The Lobster Man

By Jerry Howard

Captain George Whidden started lobstering at the age of nine in Casco Bay, Maine. Itchy for deeper waters, he moved his young family to Rhode Island in 1967. Since then he's put to sea on a 78-foot offshore lobster boat out of Galilee, fishing off the Continental Shelf in 150 fathoms with a crew of four and 1,800 traps.

But he's not fishing now. He's running—as organizer of the Coalition to Cease Ocean Dumping.

"I'm no dolphin lover or flag waver," says Whidden. "I got into this for fear of my own livelihood goin' to pieces. As I got deeper, I realized this is more than just a fishermen's issue or an economic problem. It's an American issue. We're poisoning that ocean. If we don't stop, we're gonna destroy ourselves."

In 1987, Whidden says, there were sudden drops in catch sizes, lesions on fin fish, diseased shellfish, "dolphin goin' belly up, and one dead whale with skin peeled back like sunburn, and not for lack of Coppertone."

When he visited New Jersey, fishermen told him about poor catches, large fish kills, contaminated swimming waters. Most believed the source to be a site 12 miles offshore where New York City and northern New Jersey sewage sludge and other toxic wastes had been dumped for years. Because of this, Whidden learned, the Environmental Protection Agency was

moving the dump to a new area 100 miles out to sea, 80 miles southwest of his primary lobster ground.

Capt. George Whidden of Maine is the organizer of the Coalition to Cease Ocean Dumping. He says the government is not doing its job of protecting our seas. (Jerry Howard/Positive Images)

*T*he ocean has been good to me. It's not like an oil well that runs dry. It gives and gives.

This area is one of the Atlantic's richest spawning and nursery grounds. It's also renowned for its powerful, dispersive currents, which fishermen believe are distributing poisons to fishing grounds far afield. "The only guy who really knows how tides work out there is the Guy Upstairs," says Whidden.

The coalition claims Environmental Protection Agency testing and enforcement are inadequate, that city politics has a heavy hand in all this, and that there's no time to wait for pending bills in Congress.

"The ocean has been good to me," Whidden says. "It's not like an oil well that runs dry. It gives and gives; it's like a garden down there. Those bums don't know what they're doing to it. I can't see leaving a cesspool to my grandchildren or anyone else's."

From *Modern Maturity,* August/September 1989. Reprinted by permission.

Alaska Gets a Bath—
Stone by Stone

By Pat Bodnar and Michael Cahill

"The only pay we get is satisfaction, and we manage to scrape a little bit of that out along with the oil we're cleaning up," says Jim James, a member of the Homer Area Recovery Coalition (HARC). This group of volunteers has banded together with people across the country to clean up Mars Cove—an oil-soaked wilderness area in Alaska.

Mars Cove is located 40 miles east of Homer and 300 miles south of where the *Exxon Valdez* ran aground in Prince William Sound last March [1989]. After six months of an Exxon cleanup effort, more than half of the 11 million gallons of oil spilled from the tanker remains stuck to rocks or embedded in the subsurface of Alaska's beaches.

Exxon's cleanup ended last month [September 1989], and the company has made no promises to resume work. HARC is frantically working to clean up before winter sets in. The group hopes to finish restoring the cove, stone by stone, by the end of this month [October 1989].

Nestled among Alaska's mountains and glaciers, Mars Cove is accessible only by air and skiff. About a dozen volunteers come in and out each week.

"I grew up near this area and I thought, Here's my chance to make up for some of the gasoline I've bought during my life," says James.

Jerry Brennerman, a fisherman and Alaska native, also joined the recovery effort. "The work here is a statement. It's a pretty modest effort compared to Exxon's cleanup work, but this is truly a worthwhile thing to do," he says.

The volunteers use a simple but effective technology to clean the beaches. First they set up specially designed, absorbent, recyclable booms to soak up surface oil. Then they remove shoreline rocks in buckets and run the oiled stones through a washer. Finally, they clean them by hand with a common floor scrubber.

"In our project the only thing left is the oil," says Bill Day, who designed the rock washer with fellow Homer resident Benn Levine. Day says the goal of the Mars Cove cleaning project is "to perfect the cleaning process and educate the public about the consequences of an oil spill." Day says that about $15,000 worth of equipment was donated to the Mars Cove project.

The rock washer, which looks like a giant barbecue, is considered environmentally sound but cleans very slowly. No chemicals are used. Oil and water enter the tank from the cove, and the mixture passes through packs designed to separate the oil from the water. The oil is forced to the surface of the packs, while the water continues through the washer. A generator heats the water to about 120°F. The water is then forced through a simple hose sprayer at high pressure onto the rocks. Five gallons of stones can be cleaned at a time.

Once the oil is collected, it is sent to either a hazardous waste incinerator on Alaska's North Slope or a hazardous waste landfill near Portland, Oregon.

"As we shovel this oily, mucky, terrible stuff from the cove, it just makes me more and more mad at Exxon," says Brennerman.

Jack Briscoe, a furniture builder and former school principal from Maine, says the situation is worse than he expected. "The environment's been changed and we won't know for many years all the ways that nature's been affected by the spill."

Unlike the HARC volunteers, who remove the rocks to clean them, Exxon blasted beaches with a mixture of hot water and solvents. This cleans the rock and beach surfaces, but leaves the oil beneath untouched. The company hopes beaches doused with chemical fertilizers to promote the growth of oil-eating bacteria will even-tually consume the remaining oil. But it's too early to tell if this process will be effective.

Levine says, "Given the time frame we've set for ourselves, we can clean one 70- by 30-square-foot area of the Mars Cove beach by late October. There's no good, clean, fast way to remove oil from a beach. The bottom line is not to put it in the water in the first place."

After the volunteers remove the estimated 200 to 300 gallons of oil, they will plan for spring cleanup work wherever it's needed. Organizers hope that state and federal money, combined with private contributions, will yield about $20,000 for equipment.

Coincidentally, as the volunteers rush to finish their work, the state of Alaska has just mailed out oil dividend checks. The state makes huge amounts of money off big oil, and it divides the interest among Alaskans. This year, each resident will receive a check for $873.16.

Ocean Cleanup Teaches Lessons

By Rushworth M. Kidder

Theo Keller found part of an answering machine. Bill Kinnane, his 11-year-old classmate, picked up milk cartons, napkins, "and lots of beer cans." Celina Morgan-Standard found so many disposable diapers that she wants people to start using cloth ones again.

Several years ago, when these students from the Riley School in Glen Cove, Maine, joined hundreds of others from around the state for a coastal cleanup day, they had no trouble filling 20 bags with trash from a nearby beach. And they came away with some clear convictions about ocean dumping.

"I never really litter," says Amy Halvorson, "but if I did I wouldn't do it any more."

"I don't see how they can do that!" says Bill Kinnane, talking about garbage barges that routinely dump city waste in the ocean. "They know what's going to happen to all the animals."

This year [1989], these students are gearing up for the next round: the annual Gulf of Maine Coastal Cleanup on September 23, which expects to turn out some 7,000 volunteers in Maine, New Hampshire, and Massachusetts—and, for the first time, the Canadian provinces of New Brunswick and Nova Scotia. Here in Maine, the cleanup is part of Coastweek '89, a three-week celebration observed by more than 30 states that focuses on the value of the nation's coasts.

The cleanup has had "great impact," says Kathryn Jennings, head teacher at the Riley School. Her students, she notes, are "very much aware of what litter can do, and they are very aware of what *they* can do."

And that, says Flis Schauffler, is exactly what the cleanup is intended to do. "There are so many environmental issues today that are of a scale that people just can't get a grasp on, like tropical deforestation or global warming," says Schauffler, who is communications coordinator for the Maine Coastal Program in the State Planning Office in Augusta. "There isn't that much that kids in a classroom can do. This offers them something where they can learn about the resource and really make a difference in terms of hands-on stewardship."

Richard H. Silkman, director of the planning office, agrees. "I don't think it's necessarily to clean up the shore," he says. "I think it's to draw attention to the fact that ocean dumping—trash—ends up someplace and can cause substantial damage. It's really to focus people's attention on the fact that this is a major issue for us."

For that reason, he says, the bags of litter aren't as important as the 8½- by 14-inch cards labeled "Items Collected" that each beachcombing team tallies up. The cards list 65 different types of debris, from beverage bottles and egg cartons to 55-gallon drums and lobster traps. Volunteers record each item—and use slots labeled "Other" for the refrigerators, engine blocks, bedsprings, kitchen sinks, diamond necklaces, and $100 bills that have been found in past cleanups.

Tracking the Trash

In a procedure that started with the 1988 cleanup, the information from the cards goes into the National Marine Debris Database, maintained by the Washington-based Center for Marine Conservation (CMC). The center's 1988 report—based on cards submitted by more than 47,500 volunteers in 25 states and territories who picked up nearly 2 million items on 3,500 miles of United States shoreline along the oceans and the Great Lakes—found that:

➤ The Gulf of Mexico, heavily used for shipping and for offshore oil production, was the worst victim. Volunteers in Texas, for example, found 3,549 pounds of debris per mile.

➤ By contrast, the Gulf of Maine remains one of the cleanest coastal regions. Massachusetts registered 333 pounds per mile, while Maine had only 133 pounds.

➤ Nationally, the most prevalent type of debris was plastic—which is not surprising, since it often floats. Some 62 percent of the items found were plastic, followed by paper (11.8 percent), metal (11.4 percent), glass (9.5 percent), wood (2.3 percent), and cloth (1.3 percent).

➤ Not all the trash is generated by Americans. More than 1,000 "foreign label items" from 46 countries were also counted. "Much of the foreign debris found on U.S. beaches can be attributed to dumping by the international fleet of commercial ships," says the CMC report, adding that some of it also comes from cruise liners.

➤ Some of the most compelling reasons to stop ocean littering come from the animal kingdom, as volunteers reported evidence of birds and mammals caught in nets or in plastic six-pack yokes. According to National Wildlife Federation estimates, 1 million seabirds and 100,000 marine mammals die each year from ingesting or becoming entangled in marine debris.

This year's effort promises to produce even more information. In Maine alone, some 4,000 volunteers are expected to participate—including some 1,600 who have contacted Anne Sartwell of the Kennebec Girl Scout Council, located in the Portland area. Some scout troops plan to make a weekend of it, camping out on the shore before the cleanup.

The limiting factor, says Sartwell, is access to the beaches. Much of the coastline is privately owned. Even in public areas, however, groups of children need to have access to bathroom facilities, telephones (in case of emergency), and trucks to haul off the bags of trash they collect. "That pretty much limits us to the state parks and to the town-owned beaches," she says, noting that she has not received offers from private owners to open their coastlines for the day.

The information gathered about the source of the litter will help determine what steps to take. Already, says Silkman, there have been efforts to improve trash-disposal facilities at marinas. The Maine Coastal Program published a booklet, *Charting Our Course*, which is an activity guide for grades 6 through 12 on water quality in the Gulf of Maine. And a working conference on the Gulf of Maine, convened by the governors of Massachusetts, New Hampshire, and Maine and the premiers of New Brunswick and Nova Scotia, was held December 10 through 12 [1988] in Portland.

"In the same way we went through the 1960s with an anti-litter campaign for the highways—basically we've got to go through that same exercise on the ocean," says Silkman. "There is this sense that things sink and go to the bottom," he says, when in fact much of the debris comes ashore.

"This is a preventive program," says David Keeley, director of the Maine Coastal Program, who hopes to stimulate political action to protect the Gulf of Maine while it is still healthy. "How do you get people to respond when there's not a crisis? Getting people to think about it is one of the first steps."

EARTH CARE ACTION

The Sea Turtles Come Back

By A. Rodney Tilley and Kutlay Keco

While Europe's playground—the Mediterranean Sea—is being called the "dead sea" because of its declining aquatic life, one piece of good news is emerging. Sea turtles are alive and well on Cyprus.

Turtle Hideaways

Mediterranean sea turtles, especially the loggerhead (*Caretta caretta*) and the green turtle (*Chelonia mydas*), have moved in to breed and nest on the beaches of northern Cyprus. Sea turtles once were thought to breed in concentration only on the beaches of the Greek island of Zante (Zakinthos), where about 600 turtles nest annually, and in much smaller numbers in Israel, Turkey, and Egypt.

A precise turtle count by a Cypriot environmental protection group reveals that 1,200 loggerheads and green turtles are breeding along the north and west coasts of northern Cyprus. The loggerheads congregate on Ten-and-a-Half-Mile Beach, and green turtles along the beaches of the Karpaz region. The environmentalists have sited 47 beaches or nesting sites on Northern Cyprus and targeted them for sea-turtle protection efforts.

Breeding Habits

Little was known about the breeding habits of loggerhead and green turtles in Cyprus until this decade, when local residents got involved. Through their efforts, the pattern is coming into focus. The turtles usually breed from May to September. They crawl onto the sand nightly between 11:00 P.M. and 2:00 A.M., lay between 80 and 200 eggs, and retreat into the sea before dawn. The eggs incubate for 33 to 35 days, depending on the moisture content of the sand, daily temperatures, and changes in weather conditions.

The turtles' nests are threatened by wind and wave erosion, decreasing the baby turtles' survival chances. Hatchlings often become disoriented by onshore lights and never make it to the sea. If they do not turn away from the beach lights and reach the water by dawn, the turtles will die of exposure and desiccation. Once in the sea, the young turtles have a number of natural enemies. Only 5 to 10 percent of them will live to see the next summer, and after that, chances are still low for their continued survival.

Females mature to reproduce in about seven years. Once fertile, they lay their eggs in unusual cycles—sometimes every other year, sometimes every two to three years. Some of the females, though, will lay eggs two or three times during an active year. No one knows exactly the life span of these huge turtles, but those that frequent northern Cyprus follow a typical age-related pattern of migration. The youngest females arrive in early summer and the older females come later. Female sea turtles weigh between 60 and 200 kilograms, while males are seldom weighed or tagged, unless caught by local fishermen.

Conservation on Cyprus

What brings the turtles in such numbers to northern Cyprus? It is probably a combination of instinct and environmental factors. Sea gulls, one of the newborn turtles' major predators, are nonexistent on northern Cyprus. This enhances the young ones' chances in their race to the sea and enables a larger percentage to make the water's edge. Once there, however, crabs and fishes may impede their survival.

Humans are the most active predators on sea turtles in many locations, where they have eaten turtles almost to extinction. Cypriots do not use either turtle eggs or turtle meat in their national dishes. Most other Mediterranean natives view the turtle as a gourmet delicacy.

Cyprus has far less tourism than other Mediterranean countries, but tourism is expected to rise. The northern Cyprus fisheries department has already acted to protect turtles from the encroachment of tourists. It has banned turtle killing and the disturbance of nesting sites. Tourist accommodations are allowed only up to 300 yards of the beach. Outside lighting, which also disturbs the turtles' nesting patterns, must be subdued, indirect, and at an intensity directed by environmental studies and by government decrees.

Today, ironically, the main thing that saves the Cyprus turtles from humans is that humans do not know how many there are or where they nest. Knowledge of tur-tle populations and locations could lead to destruction of these species. However, the local Cypriots are planning and acting to save their newly discovered turtle populations. They are cleaning the beaches of plastic and other debris, marking nesting sites, systematically counting the turtles, and pressing the government for even more sea-turtle protection. Their activities have led to positive public sentiment about the value of sea life, and especially endangered species. The Cypriots may be in the midst of assisting in a major Mediterranean turtle comeback.

From *Sea Frontiers*, March/April 1990. Reprinted by permission.

OZONE

While a deficit of ozone is a problem in the upper atmosphere, in large urban areas beset by huge amounts of slow-moving traffic, ozone mixes with other pollutants to create visible, dirty smog. (Otto C. Schiffert)

UPSTAIRS, DOWNSTAIRS: THE OZONE DILEMMA

By Barbara S. Hogan and Barbara Allen

Ozone is a gas that has been making headlines the past few years—alarming and, to a certain extent, confusing headlines, because some say there is too much ozone, and others say that there is too little.

A closer look, however, shows that the news is not contradictory. Where ozone is concerned, the *where* of the story is all-important.

Ozone is a highly reactive gas made up of three oxygen atoms (O_3). Unlike its cousin common oxygen (O_2), ozone occurs in only small amounts in our atmosphere and is most plentiful at two levels: near the ground and in the high-altitude portion of the atmosphere known as the stratosphere.

Downstairs, near the ground, ozone is an air pollutant that results from man-made emissions. Upstairs, however, some 10 to 30 miles up in the stratosphere, ozone is found in its greatest atmospheric concentrations and performs an important protective function: It filters out dangerous ultraviolet radiation from sunlight.

Earth's atmosphere is a vast chemical

While the stratospheric ozone layer is being depleted, the atmosphere near ground level has a problem that is just the reverse— ozone buildup near major population centers.

factory that is continuously forming and destroying substances, including ozone. When everything is working properly, a natural balance between ozone formation and destruction maintains the protective stratospheric ozone layer. It also prevents ground-level ozone from accumulating to concentrations that can harm human health and the living environment.

In recent years, however, human activities have begun to distort this all-important ozone balance. Stratospheric ozone is being depleted, while at ground level, ozone sometimes builds up to noxious concentrations near major urban areas. And there is no way to move excess ground-level ozone to the upper atmosphere, where it could be useful.

Upstairs, Ozone Loss

Recent measurements show that the amount of ozone in the stratosphere is decreasing—since 1969, by approximately 2 percent. Furthermore, researchers have reported finding a seasonal "hole" over Antarctica and some populated regions of the Southern Hemisphere, where the ozone layer thins at times by as much as 60 or 70 percent.

As the stratospheric ozone layer thins, the earth receives more of the sun's damaging ultraviolet rays. Scientists estimate that for every 1 percent decrease in the ozone layer, ultraviolet light intensity at the earth's surface increases by 2 percent. Known effects of ultraviolet exposure include greater incidence of skin cancer and eye damage among humans, and diminished crop yields for foods such as peas, beans, squash, and soybeans. Phytoplankton, the tiny one-celled plants that are a food staple for squid, fish, seals, penguins, and whales, are also vulnerable.

Where the Ozone Goes

The stratospheric ozone layer filters out dangerous ultraviolet radiation by means of a continuous cycle in which oxygen and ozone break down and re-form, absorbing ultraviolet light and releasing less damaging kinds of energy. Without enough ozone, the effectiveness of the filter decreases and more ultraviolet radiation reaches Earth.

The culprits destroying the stratospheric ozone are chemicals known as chlorofluorocarbons (CFCs) and halons. CFCs and halons are produced and released into the atmosphere by human activities and are very stable, remaining unchanged in air for as long as a century.

During their long lifetime, the CFCs and halons drift up to the stratosphere, where they are exposed to high-intensity ultraviolet radiation. Under these conditions, even stable chemicals break apart, CFCs releasing highly reactive atoms of chlorine and halons releasing bromine atoms. At this point, the array of reactions becomes bewildering, as chlorine and bromine atoms take part in multiple reactions that break ozone apart and tie up the oxygen atoms that would otherwise form new ozone. Scientists estimate that 1 chlorine atom can participate in the destruction of some 100,000 ozone molecules before the chlorine is finally captured by a reaction with hydrogen; 1 bromine atom can destroy nearly ten times as much ozone as a chlorine atom.

Even if production of CFCs and halons ceased tomorrow, the full destructive effect on the ozone shield would still be years in the future. Several decades' releases of the long-lived chemicals are making their way up toward the stratosphere—worldwide CFC and halon production peaked in 1974 at nearly a million tons per year, declined slightly during the later 1970s, and today is growing again toward the million-ton mark. Scientists project that stratospheric ozone could be depleted by between 5 and 9 percent during the next century, with an accompanying increase in ultraviolet exposure at the earth's surface of 10 to 18 percent.

Downstairs, Ozone Excess

While the stratospheric ozone layer is being depleted, the atmosphere near ground level has a problem that is just the reverse—ozone buildup near major population centers.

Near the earth's surface, small amounts of ozone are continuously created and destroyed in ongoing natural cycles. When nitrogen oxides, pollutants formed from the burning of fossil fuels, are added to the air, more ozone is created but then is quickly destroyed again.

The balance of ozone creation and destruction shifts drastically, however, when hydrocarbons are added to the atmospheric mix. Hydrocarbons—vapors from solvents, gasoline, dry-cleaning fluids and hundreds of other common substances—can actually add to the creation phase of the ozone cycle, while circumventing the destruction phase. Summer heat and sunlight speed the rate of these chemical reactions, triggering very rapid accumulation of ozone. This is why ozone bulletins are often a feature of summer weather reports but are never discussed in winter.

High ozone concentrations irritate nasal, throat, and bronchial tissues. Ozone attacks components of the body's defense system, raising concerns about the possible effects of ozone exposure on the human immune system. Ozone overdose harms forests, lowers crop yields, and damages materials such as rubber, plastics, synthetic fibers, dyes, and paints.

While ozone formation occurs most commonly over cities with large numbers of industries, power plants, and vehicles, ozone pollution is also found in remote locations—deep in the Maine woods, in rural Vermont, and on top of New York's Whiteface Mountain. Hydrocarbons, nitrogen oxides, and ozone itself can be carried by wind many hundreds of miles from their city origins, polluting rural areas.

In large urban areas where industries flourish and cars and trucks abound, ozone mixes with other pollutants to create the air pollution condition known as smog. Smog reduces visibility and irritates and inflames eye tissues.

Both ozone and smog levels peak during the summer, when plenty of intense sunlight and heat are at hand to accelerate chemical reactions. In 1988, smog reports were so bad that 96 areas across the country exceeded the federal health standard for ozone pollution, 28 more than the previous year. Nationally, 1988 ozone levels were the highest in a decade. A federal survey of 82 cities showed that in 1988, 25 percent had ozone levels higher than in any other year in the report. Unusually hot and dry weather, which fosters smog production, contributed to these record excesses.

Ozone and the High-Tech Lifestyle

Both ozone depletion upstairs and ozone excess downstairs result directly from the way we live.

CFCs that destroy the stratospheric ozone layer are widely used in industry and in consumer products—in solvents to clean electronic components, as refrigerants, as sterilants for medical instruments, and as foaming agents in fabricating a multitude of products such as pillows, mattresses, packaging, foams used in home in-sulation, and the ubiquitous food cartons and cups. CFC-foamed packaging helps protect electronic equipment, fine china, and numerous other delicate products dur-ing shipping. Halons are used primarily in fire extinguishers. The U.S. military ser-vices are major users, employing CFCs and halons to cool, protect, and clean electron-ics and weapons systems.

*T*oday America has more cars than ever on the road. In 1988, Americans owned 183 million automobiles, compared to 147 million in 1977.

Motor vehicles produce both nitrogen oxides, created when fuel is burned, and hydrocarbons, emitted when fuels fail to burn completely and when gas tanks are filled.

Today America has more cars than ever on the road, thanks to a growing afflu-ent population. In 1988 Americans owned 183 million automobiles, as compared to 147 million in 1977; the number of trucks mushroomed by 40 percent. In addition, petroleum helps us with our chores by powering lawn mowers, chain saws, and Weed Whackers, and helps us have fun by providing recreation in boats and other ve-hicles.

Hydrocarbons evaporate into the air from other sources, as well. Solvents evap-orate into the air from the millions of gal-lons of paints and other protective coatings applied each year, and during processes such as printing and chemical manufac-ture, which rely heavily on solvents.

Industry and electric power produc-tion belch out billions of tons of emissions each year, including vast amounts of nitro-gen oxides. Worldwide, fossil-fuel burn-ing, a major source of nitrogen oxides, has tripled in the past four decades.

Solutions Require Lifestyle Changes

Eliminating ground-level ozone pollution and healing the protecting stratospheric ozone shield will necessitate changes in the materials used in consumer products, in fuels that drive cars and tools, in engineer-ing design for cleaner, more efficient fuel combustion, in energy sources that power factories and homes, and in our individual lifestyles. Changes include:

➤ Replacing CFCs and halons with chemicals that do not destroy ozone.
➤ Using solvents and paints that contain fewer pollutants and fuels that burn cleaner.
➤ Finding substitutes for fossil fuels, including energy sources such as solar, nuclear, hydro, and wind power.
➤ Modifying individual habits, cut-ting out unnecessary automobile driving, and making more use of mass transit systems and car pools.

Government Takes Action

As early as 1976, concern for environmen-tal damage from CFCs caused New York State to alert aerosol spray can users to the potential danger of CFC discharges to the air. Two years later, growing public apprehension about stratospheric ozone depletion spurred the federal government to ban CFC use in most aerosol spray cans.

And in the spring of 1989, New York, along with Vermont and New Jersey, sponsored a governors' conference on global climate change. The conference report recommends that, by the year 2000, CFC production and use be eliminated and new-car fuel economy be improved to at least 42 miles per gallon.

To help combat ground-level ozone pollution, New York State requires that low-volatility gasoline (gasoline that evaporates less readily) be marketed during the summer months. For the New York City metropolitan area, the state requires that

gasoline vapors be recovered during refueling at service stations and that vehicles be inspected annually to ensure that tailpipe emissions meet established standards. Furthermore, New York State, along with seven other Northeast states, will adopt tougher motor vehicle emissions standards for hydrocarbons, nitrogen oxides, and carbon monoxide beginning with the 1993 models. The state is also assessing whether alternative motor vehicle fuels, such as methanol, will reduce ozone production.

Global Concern for the Ozone Shield

Nations are joining together to combat destruction of the stratospheric ozone shield, resulting in landmark international commitments.

In 1987, 32 countries signed a protocol in Montreal agreeing to cut CFC production in half by the year 2000. In 1989, the Montreal Protocol became obsolete when 81 countries, including the United States, pledged to end production and use of CFCs by the turn of the century. In addition, the Federation of European Aerosol Manufacturers agreed to reduce CFC use in aerosol cans by 90 percent.

There is a clear need to limit the growth of harmful emissions into the air. While government actions are important in limiting pollution, cooperation from individuals is vital to make the future a cleaner and healthier place to live.

From *The Conservationist*, November/December 1989. Reprinted by permission.

WHAT YOU CAN DO

Greatly reduce or eliminate the use of your air conditioner. A major source of chlorofluorocarbon emissions in America is car air conditioners. The installation of an auto air conditioner causes a release of 2.5 pounds of CFCs. Annual air conditioner recharges cause another pound of CFCs to be released. And there are an estimated 95 million automobile air conditioners in use in the United States: Additionally, when in use, the air conditioner reduces fuel economy by as much as 2.5 miles per gallon.

Greenhouse Crisis Foundation

The Homemaker Who Roared

By Lisa Drew

The refrigerator repairman in Lynda Draper's kitchen last winter was just doing his job. A faulty rotary compressor needed replacing, and General Electric had sent him to do the work free of charge. Everything was routine until he asked the Maryland homemaker to open the window while he vented some coolant. "I can't believe you're doing this," she said as the gas hissed out, and she proceeded to give him a fast lesson in Ozone Destruction 101.

How dare he casually release gases that contain ozone-eroding chlorofluorocarbons (CFCs). How could GE condone an action that not only weakens our stratospheric ozone layer, subjecting the planet to harmful ultraviolet rays, but contributes to the greenhouse effect? "I was absolutely flabbergasted," she recalls. "I'm sure he thought I was crazy." Little did the man know that his difficult customer would become both a major force behind environmental concessions by GE and a symbol of consumer awareness helping prompt other companies to change their ways.

It was only 4 ounces of chemicals, he explained, and a usual practice during repairs. But those answers weren't good enough for Draper, and she pursued better ones with extraordinary diligence. A GE spokesman pointed out that it was all legal. Inventors of recycling equipment confirmed that technology exists to capture the gases. Environmental groups shared Drap-

Lynda Draper took on General Electric over CFCs released in her own kitchen. (Art Stein)

er's outrage and encouraged her. Finally, she ended up face-to-face with five GE executives in the office of Sen. Albert Gore, Jr., of Tennessee, one of the ozone layer's champions.

Gore served as mediator and Draper as a consumer advocate threatening to blacken GE's reputation. Last August

[1989], GE agreed to offset its emissions of as many CFCs as it is releasing in the $450-million refrigerator repair program. In other words, GE will reclaim or conserve 300,000 pounds of CFCs from elsewhere in the company—if not the same gases hissing from more than a million refrigerators. "It's not quite the same thing as not putting the stuff out there in the first place," says Draper. "But there are no laws in place; they didn't have to do a thing."

A Ghastly Enemy Begins to Retreat

Other corporations have responded to consumer concern and signs of impending government regulation with similar actions. American Telephone & Telegraph

will stop using CFCs in manufacturing by 1994, halving its use by 1991. By 1991, General Motors dealerships will recycle CFCs from car air conditioners under repair. Nissan will completely replace CFCs in its air conditioners by 1993. Du Pont plans to spend $24 million looking for CFC alternatives.

Although these efforts may not be having much effect yet, to Draper they're at least tangible evidence of a fight against a ghostly enemy. "You can't see it," she says, "you can't feel it, you can't touch it." But you can teach about it—whether to an executive or a repairman. "The more people we can educate about what's going on," says Draper, "the better."

From *National Wildlife*, February/March 1990. Reprinted by permission.

EARTH CARE ACTION

Firms Cooperate to End CFC Use

By Ron Scherer

Kenneth Taulbee, an engineer from White Consolidated Industries, Inc., is being quizzed at a conference about a compressor used in a refrigerator made by his company.

Taulbee glances at the name tag of the inquisitor, who is from General Electric Co.

"Hey, we're working with you guys," says Taulbee.

In normal times, Taulbee would have considered the information none of GE's business. Or the conversation might have sparked the interest of the antitrust division of the Justice Department, since White

Consolidated and GE compete against each other on the appliance-showroom floor.

But these are not normal times.

Instead, similar conversations are taking place all over the country as companies that use chlorofluorocarbons (CFCs) are co-operating instead of competing with one another to find alternatives to CFCs, which deplete the earth's ozone layer.

Taulbee's conversation took place this month [October 1989] at a conference on CFCs in Washington, D.C.

There are now at least eight consortia or trade associations exchanging data that in the past was considered proprietary information. These groups are from such industries as refrigeration, air conditioning, foam production, fire fighting, electronics, and defense.

"Cooperative America"

"This is cooperative America, not corporate America," says Stephen Seidel, an official at the Environmental Protection Agency (EPA), which sponsored the conference. "It is unusual in the pollution field," says David Doniger, director of the Natural Resources Defense Council's (NRDC) ozone protection project.

An example of the cooperation is the refrigerator manufacturers. They are now sending each producer a compressor that will be used to calibrate test equipment. Then each manufacturer—with the knowledge of its competition—will test a different CFC alternative. They will compare results.

Since refrigerators usually last 15 to 20 years, the testing has to evaluate long-term

effects. "This will reduce the amount of testing by each company," says Jan Michael Pottinger, a senior staff engineer with Admiral Home Appliances, a division of the Maytag Corp., in Galesburg, Illinois.

Defense contractors, who normally compete against one another, are also banding together. Such companies as Texas Instruments, Inc., Northern Telcom, American Telephone & Telegraph Co., and Boeing Co. are sitting down with the Department of Defense and rewriting the military specifications that apply to CFCs.

The old military specification required companies either to use CFCs or apply for permission to use an alternative. The new specification allows the companies to use an alternative without seeking permission as long as the contractor includes technical analysis of the new process.

At the same time, the defense electronics contractors are working together to evaluate alternative products.

This who's who of defense suppliers is comparing data on solvents used to take the grease and metal shavings from electronic circuit boards.

"If you participate, you are required to share the information," says Joe Felty, manager of process information for Texas Instruments. Earlier this month, seven large companies, including Texas Instruments, pledged to both phase out the use of CFCs by the end of 1994 and to share information.

Competing companies have worked together on a smaller scale before. The semiconductor companies, with government funding, have formed a research and development cooperative, SEMATECH, to help compete against the Japanese. Trade associations routinely put together technical committees to establish industrywide standards.

According to the Justice Department, there have been 130 joint research projects—ranging from the steel industry to the auto industry—begun under the auspices of the National Cooperative Research Act of 1984. This act limits the potential liability of companies from antitrust prosecution.

However, as James Calm, director of the Air Conditioning and Refrigeration Institute notes, "This is the first time the government has sought to be the catalyst."

EPA officials agree. "We were driven most by the tight time frame," Seidel says. The United States and 40 other nations signed the Montreal Protocol, which scales back CFC use to 1986 levels by this past June [1989] and gradually scales back all use to 50 percent by the year 2000.

The United States and Canada, the European Community, and the Nordic countries are calling for a complete phaseout of CFCs and halons (used in fire fighting) by the end of the century.

In the meantime, many companies are setting their own timetables for the elimination of CFCs, and some states and communities have set deadlines. Congress is also considering legislation mandating CFC cutbacks.

Global Cooperation

The cooperation goes beyond U.S. borders. "All data are available to every industry in every country to accelerate the phaseout of CFCs," explains Calm. There has to be such global cooperation because of the way companies source parts for machines. Many refrigerator companies, for example,

WHAT YOU CAN DO

Encourage your local auto service center to install and use CFC recycling equipment for auto air conditioning repair. There are inexpensive machines available that capture and recycle the CFCs released during air conditioner service and repair.

Greenhouse Crisis Foundation

use compressors built in Japan by Matsushita. And electronic circuit boards are produced in scores of countries around the world.

Despite the cooperation, EPA officials maintain companies will remain competitive. "People will still want to be first and best," Seidel says.

Using the same materials in manufacturing, says Felty, does not necessarily mean the products will come out looking the same. "The processes vary so greatly that it does not mean there will be a consolidation of designs," he says.

David Doniger of the NRDC says there has been no evidence that the consortia are other than earnest. "This is not just public relations," he says.

That much is clear from the conference. In working sessions, competitors compared notes on different chemicals. Almost every speaker offered to send technical information to anyone sending in a business card. "There is a lot more creative thinking on the issue," says Doniger.

Old-fashioned crop dusters still spread lethal pesticides, but they can also be used to spray alternative pest controls, such as the larvae of the green lacewing, which attack a variety of harmful insects. (Ed Landrock/Rodale Stock Images)

AMERICA TACKLES THE PESTICIDE CRISIS

By Susan Gilbert

Jim Knutzon is sitting in the restaurant of the Hilton Hotel in Pleasanton, California, near Oakland, solemnly regarding the plate of swordfish smothered in sauteed vegetables just set before him. With head bowed, he delivers what might be construed as a modern-day meditation. Knutzon, a former pesticide salesman who now advises farmers on pesticide alternatives, expresses the hope that people across America will soon be able to sit down to meals that are free of toxic pesticide residues. This said, he begins to eat.

Knutzon's wish is not unrealistic. There are powerful stirrings throughout the country, from the vast fruit and vegetable farms of California to the cornfields of Iowa to the apple orchards of New England, not to mention Capitol Hill, indicating that America is making substantial changes in the way it grows its food. Tens of thousands of farmers have scaled down their use of pesticides they relied on for decades—compounds that in many cases were not adequately tested for safety and are now known to be dangerous to human health. Major supermarket chains and food manufacturers are refusing to buy produce that contains even minute, legally permissible residues of certain pesticides. And Congress is considering a bill that would, in all likelihood, restrict the use of dozens of pesticides to such an extent that they would no longer be profitable to produce.

In lessening the nation's dependence on agricultural chemicals, farmers, food manufacturers, supermarkets, and legislators are all taking their cue from the top of the food chain: the American consumer. People are now well aware that good nutrition and all-around healthful behavior can work in tandem to help spare them from many of the illnesses that claimed their parents and grandparents. But the increasing evidence that pesticide residues in foods can cause cancer and other serious health problems has transformed the private matter of eating well into a political issue with a resounding message: Get hazardous pesticides out of the food supply, and do it now.

Farmers on the Front Lines

As Jim Knutzon sees it, farmers have been twice victimized—first by the government and the chemical companies, which lead them to believe that pesticides were safe, and second by the public, which now expects farmers to give up pesticides overnight. "Still, more and more farmers are saying, 'If consumers are lit up about pesticides, I ought to be a thousand times more lit up,'" he says.

Knutzon recently joined NutriClean, an Oakland company whose goal is to get most pesticides off the farm and out of the American diet. Toward this end, the company serves as a consultant to both supermarkets and growers. Among the alternatives to pesticides that NutriClean recommends to farmers is the release of insects that are actually beneficial because they prey on other insects before those bugs can prey on a crop. Ladybugs, for example, eat the eggs of most of the worms that eat lettuce and cabbage.

*T*he trend away from agricultural chemicals is strongest in California. The number of organic farms there, which use no man-made pesticides or fertilizers, has doubled in the past year.

The trend away from agricultural chemicals is strongest in California, where about 53 percent of America's vegetables and 42 percent of its fruits and nuts are grown. In California the number of organic farms, which use no man-made pesticides or fertilizers, has doubled in the past year, to about 1,500. In addition, thousands of other California farms now use pesticides—a term that encompasses insecticides, herbicides, and fungicides—only as a last resort. Together, these farms grow and ship to the rest of the country virtually every kind of fruit and vegetable that Americans eat regularly: lettuce, tomatoes, celery, carrots, broccoli, oranges, peaches, grapes.

Driving his white Mercedes on the dirt roads that surround his farm, Don Goldberg, a grape grower in Kelano, California, in the San Joaquin Valley, proudly points out the differences between his field and that of his neighbor, the Dole Food Co. Goldberg, who is co-owner of the Skyline Ducor Ranch, has been able to reduce his pesticide use by 78 percent in the past three years, chiefly by releasing beneficial insects onto his 400 acres. Larvae of such insects as the green lacewing are literally sprayed onto the field either by a crop duster or by a rig fitted on top of a tractor. These bugs devour mealybugs, thrips, worms, mites, and other pests that can devastate a grape crop.

The soil between the rows of Dole's grapes is completely free of weeds, a sign that herbicides are routinely sprayed. By contrast, the soil between the rows of Goldberg's grapes is barely visible under a covering of tall grasses and flowers. Those plants provide three things: a breeding ground for beneficial insects; nitrogen for the soil, which in turn reduces the need for fertilizer; and a lure for harmful insects that would otherwise attack the grapes.

"This is a holistic system," says Goldberg, a trim, athletic-looking man with

It looks like something from outer space, but the "Beetle Eater" is a giant vacuum that sucks up insect pests from potatoes, peas, peppers, and all kinds of beans. The four-row model shown here costs about $15,000. Growers say it is more effective, safer, and cheaper than insecticides. The Beetle Eater is sold through Thomas Equipment, P.O. Box 130, Hawkins Road, Centerville, New Brunswick, Canada, E0J 1H0 (506) 276–4511. (Mitch Mandel/Rodale Stock Images)

white hair and bronzed skin. His point is that no one technique can serve as a substitute for pesticides.

Even something as simple as redesigning his grape trellises has helped Goldberg reduce the amount of pesticides he must use. He has found that by making the trel-

lises taller and wider, more air circulates between the vines, thereby significantly decreasing the risk of fungal infections.

The transition from chemical to non-chemical farming can be costly, often requiring an initial investment of many thousands of dollars. But, as Goldberg sees it, farmers don't have a choice: "If the consumer wants a two-door car, don't try to give him a six-door car. I saw this trend coming, but it got very intense this year because of that Alar thing and that Meryl Streep."

Protecting the Children

The public was galvanized into acting against pesticides by a shocking report last February [1989] from the Natural Resources Defense Council (NRDC), an environmental group. The report concluded that children face a tremendous risk of getting cancer and other health problems, such as kidney damage and depressed immune function, from pesticide residues in fruits and vegetables. According to the council, the cancer threat to children is as much as *460 times greater* than is considered acceptable under federal law. The main reason children are so vulnerable is that they eat many times more fruit and fruit products than the average adult—on whom the government bases its safety standards.

Among the NRDC's findings: Between 5,500 and 6,200 of today's preschoolers may get cancer sometime in their life solely because of their exposure to eight pesticides in amounts typically found in fruits and vegetables—amounts that are well within federal limits. In addition, at least 3 million of today's preschool-age children are being exposed to other pesticides in fruits and vegetables at levels—again, legal levels—that are potentially harmful to the brain and other parts of the central nervous system.

The chief villain in the NRDC report was Alar, a growth regulator that, as of last spring, was sprayed on about 15 percent of the nation's apple orchards to make the fruit ripen at the same time. A number of studies had found that when the treated apples were processed—to make apple juice, for instance—Alar broke down into a potent carcinogen called unsymmetrical dimethylhydrazine, or UDMH. There was evidence, too, that Alar itself was carcinogenic.

Alar quickly became a symbol, and actress Meryl Streep spearheaded a crusade against it. In television and newspaper interviews, she urged parents to buy organic produce. And in passionate testimony before the Senate, she suggested that the government's failure to take Alar off the market amounted to little more than a laboratory experiment on the nation's children. The public responded with a vengeance. School cafeterias stopped buying apples. Supermarkets stopped doing business with Alar-using growers, and growers got stuck with warehouses full of suddenly unsaleable fruit.

As a result, Uniroyal, the manufacturer of Alar, stopped selling it in the United States last June. (The company still sells it abroad.) But that was only one step toward safer food.

As an indication of the scope of the problem that remains, the National Academy of Sciences has listed more than a dozen foods in which pesticide residues present potentially significant risks. Tomatoes top the list. Next is beef, because the grains fed to cows contain pesticides that accumulate in the animals' tissue. The other foods cited by the academy are potatoes, oranges, lettuce, apples, peaches, pork, wheat, soybeans, all other beans, carrots, chicken, corn, and grapes. The academy believes that these foods account for 95 percent of the health problems associated with pesticides.

One reason the academy considers pesticide residues so dangerous is that they can't all be washed off. Apples, tomatoes, and cucumbers, for example, are typically coated with a fungicide mixed with wax to keep them looking fresh and prevent them from spoiling. The wax makes it impossible for the fungicide to be rinsed off. And while peeling fruits and vegetables may get rid of pesticides on the skin, it doesn't remove those that are absorbed by the plants' leaves or taken up by the roots. These pesticides, known as systemics, become part of a plant's chemistry. Bananas, potatoes, and melons are among the crops often grown with a systemic called aidicarb, which even in small amounts can cause temporary nervous disorders, stomach cramps, and various other ailments in children.

Something Is Wrong with the System

The Environmental Protection Agency (EPA) does not dispute the scientific findings that legally permissible levels of pesticides can be dangerous. Indeed, the agency has estimated that pesticide residues cause 6,000 cases of cancer each year—many more than exposure to asbestos in homes and schools and as many as passive smoking. While that figure is less than 1 percent of the 900,000 cases diagnosed each year, health experts point out that cancers caused by pesticide residues are preventable. "Among the easiest cancers to prevent are those with causes that can be regulated by the government, like pesticide residues in foods," says Dr. Robert N. Hoover, an epidemiologist at the National Cancer Institute.

What many people find difficult to comprehend is why the government permits harmful pesticides to stay on the market. The National Academy of Sciences says the government actually understates the dangers of pesticide residues because it relies on outdated information. The EPA estimates that Americans eat about one avocado, one artichoke, one nectarine, one wedge of melon, a quarter of an eggplant, and a handful of mushrooms each year. But health-conscious individuals are likely to eat far greater amounts of these foods. Department of Agriculture statistics indicate that consumption of many fruits and vegetables has increased significantly in recent years, largely at the urging of nutritionists and health authorities such as the Surgeon General. It is ironic that, in striving to improve their health, people end up consuming more pesticides. This is not to say that people are wrong to change their diets. But the National Academy of Sciences points out that as consumption of fruits and vegetables rises, the need to restrict pesticide use becomes more crucial.

In the EPA's defense, Rick Tinsworth, an official in the agency's pesticide program, says that in many cases delays in banning chemicals are caused by pro-

tracted court hearings. (The EPA has banned only 26 pesticides since it was created in 1970, although it has said that about 70 now in use are carcinogenic in animals.) While admitting that "some cancellations of pesticides take too long," Tinsworth explains that chemical companies routinely bring lawsuits against the government for trying to ban their products, arguing that there is a lack of scientific data. He cites bureaucratic "glitches" and insufficient funds as other factors that slow down the process of banning pesticides. In 1988, the budget for reviewing the risks associated with pesticides was just under $14 million, an amount that Tinsworth maintains was inadequate to finish the task before "well into the next century."

Many agricultural and health experts believe that despite its financial constraints, the agency has been dragging its feet. "EPA has become averse to being sued and has only taken chemicals off the market that have been uncontested," says Charles M. Benbrook, executive director of the National Academy of Sciences' Board on Agriculture.

The academy has recommended that the EPA be required to accelerate its review of all 40,000 pesticide products that came on the market without proper testing. In part because of that recommendation, Congress last year [1989] passed an amendment to the pesticide law that orders the agency to complete its review by 1996 and doubles the budget for doing so. The academy has also urged Congress to simplify the law, stating, in effect, that it is so labyrinthine that even EPA bureaucrats don't completely understand it.

A bill currently before Congress would both streamline the law and make all pesticides subject to more stringent health standards. For the first time, the effects of pesticide residues on infants, children, pregnant women, and other vulnerable people would have to be taken into account. If a product has the potential to cause more than one case of cancer for every million people exposed to it, any of the following actions would occur: The product would be banned for use on particular crops; the number of times it could be applied to a crop would be reduced; or, if these restrictions were inadequate, the pesticide would be swiftly taken off the market—regardless of its benefits to agriculture. The bill, sponsored by Rep. Henry A. Waxman of California and Sen. Edward M. Kennedy of Massachusetts, could result in the biggest change in American farming since the green revolution.

F or centuries, farmers used natural poisons to kill insects, and some vegetables came to market dusted with arsenic.

Together with the invention of mechanized farm equipment and the hybridization of vigorous crops, pesticides helped bring about a lifesaving explosion in food production. Chemical fertilizers and pesticides have been instrumental in enabling farmers to grow all the food the world needs; that hunger still exists is largely a distribution problem.

Most pesticides used today are man-made compounds. The first widely used synthetic pesticide, DDT, was sprayed by the United States Army during World War

II to prevent the spread of malaria by killing mosquitoes. It was so effective that by the end of the war, American farmers had begun applying it to their crops to kill a wide range of other pests.

Contrary to popular belief, the health problems associated with pesticides did not begin with DDT. For centuries farmers relied on natural poisons such as arsenic, lead, and copper, and according to historical accounts, fruits and vegetables would occasionally come to market dusted with these pesticides.

As far back as 100 years ago, journalists were writing that the residues might be dangerous. But their warnings were largely ignored by the public because there was no scientific proof. An article in the *New York Times* on September 25, 1891, under the headline "Poisoned Grapes on Sale," reported that the city's health department had ordered that all grapes covered with a greenish powder be removed from grocery store shelves after determining that the coating on some of the grapes was a potentially toxic concentration of copper sulfate. Other cities in the East then threatened to ban the sale of all grapes. Fearing for their livelihood, a number of Hudson Valley grape growers appealed to the United States Department of Agriculture (USDA) for help. The USDA insisted that the residue was safe and criticized the health department for overreacting. In the days that followed, the paper changed its tone, referring to the incident in one headline as "That Silly Grape Raid."

At about the same time that the agriculture department was proclaiming the safety of New York State grapes, Britain and France were passing laws restricting pesticide use. In 1892, an editorial in the

British Medical Journal warned that apples and other produce imported from the United States might be poisonous, because they often arrived covered with arsenic powder. The response of American apple growers was to accuse the journal of attempting to hurt their sales.

Although the apple growers refused to abandon pesticides, many American farmers in the late nineteenth century still did not use them, either because they didn't think they needed them or because they feared being poisoned. The number of holdouts was significant enough to rattle the agriculture department, which went as far as to put pressure on farmers who didn't apply arsenic, lead, or other poisons to their fields. At the department's urging, several states passed laws requiring that all farmers use insecticides, on the theory that the insects in unsprayed fields were bound to infest neighboring farms.

After the turn of the century, medical evidence mounted that pesticide residues on fruits and vegetables were poisoning and in some cases killing people. Emotional pleas for reducing pesticide use were made in books, magazines, and newspapers. More than 50 years ago, a book titled *40,000,000 Guinea Pig Children* made an appeal strikingly similar to Meryl Streep's; the author warned that children were especially vulnerable to pesticide poisoning because they ate more fresh fruit than adults. As a result of such evidence, the federal government in 1947 passed the first law explicitly regulating the use of pesticides.

The law went into effect just as DDT was gaining in popularity. Once farmers had DDT, which worked faster and killed more insects than any chemical they'd ever used, they rapidly abandoned arsenic and other age-old pesticides. Confident that they had the ability to conquer their chief enemies, many farmers expanded their operations. Small farms with a patch of lettuce here and some tomatoes there turned into large farms with a thousand acres of nothing but lettuce. Farming became a factory-style business. And soon after DDT came on the market, other powerful pesticides emerged.

But now enter Rachel Carson. Her book *Silent Spring*, published in 1963, described as never before the threat of pesticides, conjuring up a ghostly American landscape: Birds were absent, cows had stillborn calves, and people lay dying of cancer, all because of pesticides in the air, water, and food. Carson, a biologist, supported her case with scientific evidence that DDT and other pesticides persist in the environment for decades, entering the food chain and building up in the tissues of humans and other animals; she also pointed out that babies swallow pesticides with their mothers' milk.

After the publication of *Silent Spring*, a consensus grew among scientists and government officials that pesticides as they were being used had the potential to harm human health. Federal regulations governing the testing and use of agricultural chemicals were strengthened, and in 1972, DDT was banned. (It is slow to break down, however, and is still one of the most commonly found residues in many foods grown in this country.) Despite the increasing awareness of the dangers of pesticides, the amount of chemicals deposited on American farms continued to grow.

The use of agricultural pesticides in the United States peaked in 1982, when 880 million pounds were sprayed, a tenfold increase in the nearly 40 years since synthetic pesticides were introduced. During the same period, farm productivity more than doubled. Clearly, large-scale commercial farming could not exist without any pesticides. But since 1982, their use has been declining, as more and more farmers have come to realize that many of the chemicals they once routinely applied to their fields are not really necessary.

Pesticide Alternatives Do Work

The decline in pesticide use is welcome news to people concerned about their health, but many experts take the position that as long as most farmers ignore effective alternatives to insecticides, herbicides, and fungicides, these chemicals are being overused. A report released last month [September 1989] by the National Academy of Sciences concludes that alternatives to pesticides not only work but are economically viable. The report indicates that the potential exists for an even greater transformation of American farming than the one now under way. "You don't need any of the worst chemicals," says Charles Benbrook, whose staff helped put together the report. He says that even if a significant number of pesticides were suddenly taken off the market, "there's no way that any major fruits and vegetables would become unavailable."

The alternative farming methods cited by the academy include the widespread use of disease-resistant beneficial insects and crop rotation—planting a crop in a different field each season to prevent insects, fungi, and microbes from gaining a foothold in any one area. In places where these methods would not work for particular crops, the report recommends that the crops simply not be raised. "Most fruits and vegetables probably would not be able to be grown in the Southeast, because in hot, humid climates, pests are too great a problem," says Benbrook. "Even now, the cost of pesticides is killing farmers there." He believes that virtually all of the nation's fruits and vegetables—the crops that are most vulnerable to insects and diseases—should be grown in California and other hot, dry Southwestern states, where the alternatives to pesticides work best.

Jim Hightower, Texas commissioner of agriculture, who for a number of years has initiated programs in his state to encourage farmers to adopt nonchemical practices, attests to their success. "What we've had here, and what we found elsewhere, is that farmers doing this kind of agriculture make money," he was quoted as saying a few weeks ago. "They pay off their notes, they pay for tractors, and they do it without crop subsidies. There is conversion coming."

A reduction in pesticide use sometimes means higher food prices. One need only walk the aisles of a health food store to see that many organic fruits and vegetables are selling for double or more the price of conventionally grown produce. It's true that organic farming is labor-intensive, and labor is expensive. Still, many people are willing to pay more for that they believe to be healthier food.

But what many supermarket chains around the country, including some of the largest, have begun to prove is that people don't have to pay much, if anything, more for produce with little or no pesticide residues. Several of these supermarkets, among them Stop & Shop in the Northeast and Raley's in the Sacramento area, sell fruits and vegetables that have been certified by NutriClean as having no pesticide residues detected in laboratory tests. The produce comes from farms that are not necessarily organic but that keep pesticide use to the minimum.

"Only rich people can buy organic food," says Stanley Rhodes, the president of NutriClean. "With our system we can keep food at market prices and, most important for consumers, get pesticide residues out of the food." Rhodes is a former 1960s rebel who now wears a suit and tie and sits in a large corner office overlooking the modern buildings of downtown Oakland, California. He is the son of a winemaker in the Napa Valley who refused to use grapes grown with pesticides, just like his father, his grandfather, his great-grandfather, and his great-great grandfather—all winemakers in Germany.

Rhodes, who holds a Ph.D. in chemistry, is quick to point out that he is not anti-pesticide. Some pesticides help protect the public from natural substances that are harmful; aflatoxins, for example, chemicals produced by a fungus that grows on peanuts and corn, are much more potent carcinogens than any fungicide used to destroy them. And Rhodes says that some of the toxins that plants themselves produce to fight off insects may be just as dangerous to human health as pesticides. For these reasons, he eventually would like to test food, whether it is grown conventionally or organically, for the presence of all kinds of toxic substances. But for now, his focus is on pesticides.

In addition to certifying produce, each week NutriClean takes samples of all the fruits and vegetables delivered to the supermarkets that participate in its program and tests each one for the presence of up to 200 pesticides. If any are found, NutriClean immediately notifies the supermarkets; the stores can return shipments that contain food with residues in excess of legal limits. At the end of each week, NutriClean circulates a memorandum to the markets listing those farmers whose goods showed residues that week and those whose produce didn't. The philosophy is simple: Information is power. What this really amounts to, Rhodes says, is giving consumers more power over their own food system. NutriClean's success has prompted a few other companies to jump into the food-testing business.

Organic Growing Takes Off

The fear of being blackballed by supermarkets is but one of many factors that are causing farmers across the country to cut down on pesticides. In Iowa, for example, some 300 growers of soybeans, corn, and other grains are trying alternatives to pesticides because the state government is encouraging them to do so. A law passed in 1987 caused a sharp rise in pesticide prices by increasing the fees that manufacturers must pay to register the chemicals and that retailers must pay to sell them. The fees help fund a research program that gives

At the Rodale Research Center in Kutztown, Pennsylvania, researchers work with new ways to grow crops, such as these broccoli plants, without relying on dangerous pesticides. (Rob Cardillo)

farmers information on alternative means of pest control.

Following their colleagues in California, New York, and a number of other major agricultural states where organic farms have been certified under voluntary programs for several years, New Jersey's organic farmers recently set standards for certification. New Jersey ranks in the top ten in the nation in the production of various fresh fruits and vegetables, including peaches, blueberries, and tomatoes, and the certification program is giving many conventional farmers the incentive to try organic methods.

Frank J. Stiles, a second-generation New Jersey farmer, is cultivating about 90 of his 200 acres organically this year. His organic crops include tomatoes, bell peppers, broccoli, cauliflower, melons, blueberries, asparagus, and apples. "The more you read, the more you wonder whether you need it all," he says, referring to the pesticides he relied on. "You worry about your safety and your workers' safety."

Instead of using synthetic insecticides,

Stiles uses a spray derived from a bacteria that kills worms without harming anything else. He eliminates other insects by releasing their natural predators and controls weeds with thick layers of mulch and weeding machines. To fight crop diseases, Stiles uses copper and sulfur sprays—the same pesticides that were so controversial in the past—but he applies them in smaller amounts and well in advance of harvest time to help minimize residues. None of the pesticides he uses are systemics, so they can be washed off.

So far, Stiles's results have been mixed. "The crops aren't as good as with chemical controls," says Helen Athowe, a horticulturist who manages the organic portion of the farm. "We will probably have more crop losses." While Stiles's experience is fairly typical of farmers who try to switch to organic methods, Athowe is confident that the problems they've encountered are solvable. After a few years of organic farming, growers tend to learn which methods work best on which fields, and their crops generally improve.

A Pesticide-Free Future?

In the major apple-producing states of Washington, New York, and Massachusetts, organic apple farming is extremely difficult. Apples have been grown for so long in these states that the insects and diseases that plague them are strongly established. But with the help of state and private programs aimed at reducing pesticide use, many growers have cut back on spraying without harming the quality or quantity of their fruit. In fact, in a five-year ex-

periment begun in 1978 by Ronald J. Prokopy, an entomologist at the University of Massachusetts at Amherst, 40 orchard owners were able to reduce their use of insecticides by 40 percent and bring down their costs without sacrificing yields or quality.

Alar was a crucial tool for many of these growers—those who raise McIntoshes, in particular—because it helped prevent the fruit from dropping prematurely and attracting insects that would then attack the trees. David G. Bishop, a McIntosh grower in Massachusetts and Vermont, says that without Alar he had to make one extra application of insecticides last summer. Even so, Bishop's overall pesticide use was less than it was seven years ago, when he stopped spraying all of his trees according to a fixed schedule and adopted a policy of applying pesticides only when and where they were needed.

Although consumer demand bears much of the responsibility for the recent decline in pesticide use, farmers point out that it was consumer demand that impelled them to use pesticides in the first place. Americans like uniformly red apples, unblemished tomatoes, peaches without bruises. But such cosmetic perfection can be achieved only with pesticides. It was Alar, for example, that made McIntoshes enticingly red; this year's crop is expected to be streaked with more green, although the change in color will not affect the taste.

A study by the California Public Interest Research Group concluded that 40 to 60 percent of the pesticide applications on tomatoes and 60 to 80 percent of those on oranges are made primarily for cosmetic purposes. But in another study by the group, two-thirds of the consumers surveyed said they would pass up a flawless orange in favor of one with scarred skin that was grown with half the amount of pesticides.

Consumers are beginning to realize that the best-looking produce is not necessarily the best for their health—and demanding what is. Now that farmers have

strong economic incentives to stop using so many pesticides, Charles Benbrook predicts that what is now called alternative agriculture may, in a decade or so, be mainstream.

EARTH CARE ACTION

Maine Finds Promise in Fish Wastes

By Keith Schneider

Fish-processing companies and researchers in Maine have developed techniques to turn fish wastes that have been a malodorous disposal problem into compost, animal feed, and a biological pesticide.

These products of an emerging industry to recover natural resources from garbage also have the potential to reduce pollution from chemical fertilizers and pesticides.

"We are engaged in raising the consciousness of our fishing industry," said Michael D. Moser, director of economic development at the state Department of Marine Resources in Augusta. "We don't like

to use the word 'waste.' This is a valuable material for further processing."

Spray Seems to Kill Insects

Among the most promising of the new products is a liquid spray manufactured by Biotherm International, Inc., a small biotechnology company in Portland. It is being tested weekly on potato plants at the University of Maine campus on Presque Island. The spray seems to kill weeds, provide nutrients, repress bacterial diseases, and ward off insects, including the destructive Colorado potato beetle.

"We use 28 different agricultural chemicals to grow potatoes in Aroostook County, and some of the chemicals have ended up in groundwater," said William H. Forbes, executive director of the Maine Research and Productivity Center, a state agency that is based at the campus and has invested $10,000 in Biotherm's research. "If we can use a natural product as a substitute, the benefits to the environment are clearly apparent."

Maine's 25 fish processors once hauled tons of fish wastes to two rendering plants in the state and one in Massachusetts that used it as an ingredient in animal feed. But as soybeans became the principal source of protein in feed, the demand for fish fell and the last of the rendering plants, in Rockland, closed last year.

But the need to dispose of up to 25,000 tons of fish bones, guts, heads, and tails produced by Maine's processors each year has not ended.

Several companies have been given state permission to dump the wastes at sea. Others now earn $5 a ton from a plant in Boston that turns fish wastes into high-quality feed used to raise minks in the Midwest.

Increasing the Cost of Fish

A few companies spend thousands of dollars a week shipping wastes to a Canadian fishmeal plant in Blacks Harbor, New Brunswick. "It adds 3 to 5 cents to the price of every pound of fish we sell," said Frank J. O'Hara, president of F. J. O'Hara and Sons, a fish processor in Rockland.

Recognizing the problem for one of Maine's most important industries, state officials began offering grants for promising proposals. The state's interest has led to an array of new uses for fish wastes, particularly in agriculture.

North Atlantic Products, Inc., in Rockland, a processor of dogfish, a species of shark, received permission from the state to build a plant to turn its wastes into compost. The idea is as old as the Pilgrims, who were taught by Indians to place small fish in the holes with their corn seeds.

North Atlantic's plant is to mix up to 20 tons of fish wastes daily with sawdust to provide carbon and is to be built this fall [1989] in Thomaston. John Gallagher, the company's business manager, said the company will market its new product, Seagreen, to gardeners and organic farmers.

The Stinson Canning company built a plant in Bath to turn wastes into ingredients to feed young hogs and salmon and trout on fish farms. A third company, Maine Fisheries, in Portland, is developing a process to manufacture feed pellets for aquaculture and livestock from wastes, the Department of Marine Resources said.

Cracking the Market

"The tough nut to crack in all of these projects is the market," said Michael Moser of the marine resources department.

That may not be the case for Biotherm with Biostar, the liquid compound being tested here. In its pure state, Biostar looks and smells as if it had been siphoned from a cesspool. But after being diluted in water, Biostar exhibits some remarkable traits.

Since early July a potato field on a hill overlooking Presque Isle has been sprayed weekly with Biostar. Test plots in the field that have been sprayed with a solution that is 90 parts water and 10 parts Biostar seem to repel Colorado potato beetles. Sprayed plants appear to be hardly touched by the ravenous bugs. Meanwhile, potato plants in plots that have not been treated have been eaten to the veins.

The researchers do not know precisely why it works. They have also not yet determined its possible toxicity to animals and people, nor its ability to spread into groundwater.

The sprayed plants are greener and healthier than untreated plants, which may account for the lower populations of insects that feed on the treated plants.

"It's a well-recognized principle in agriculture that bugs are attracted to plants with yellowed leaves or ones that are weakened by stress," said Spencer Apollonio, a marine biologist and consultant to Bio-

therm. "Our hypothesis is that the spray makes plants so healthy, bugs are not attracted to them."

In northern Maine, as in other important United States farming regions, the interest in natural compounds is keen. Most of the $28.4 million that Maine's farmers spent last year on 77,000 tons of chemical fertilizers and thousands of pounds of pesticides was spent in Aroostook County, which is as large as Massachusetts and is the third largest potato-producing area in the nation after Idaho and Washington State.

About 90 percent of the state's $110-million potato crop is grown in the county, and the heavy use of chemicals has taken a toll on the quality of the region's groundwater. A half-dozen insecticides and herbicides and nitrates from fertilizers have been detected in groundwater, reported the Maine Geological Survey, a state agency.

"We'd like to get away from the chemicals if we can," said Gregory Smith, the manager of Maine's largest broccoli grower, Herschel Smith Farms, in Blaine. Smith has begun testing Biostar on broccoli to control a bacterial disease and to determine its value as a fertilizer. "It's worth it to us to give this stuff a try and see what happens," Smith said.

WHAT YOU CAN DO

Birds can play an important role in minimizing garden pests. One swallow devours 1,000 leafhoppers in 12 hours—one brown thrasher can eat over 6,000 insects in one day! Create an environment attractive to birds by providing natural food sources such as berry bushes, water, suitable nesting areas, roosting sites, and protection against animals.

One Person's Impact

From the *New York Times*, August 20, 1989. Copyright © 1989 by The New York Times Company. Reprinted by permission.

Looking for Mr. Goodbug

By Dwight Holing

It's harvest time in Arizona's lush Harquahala Valley, and grower Stephen Pavich is inspecting his vines. The shoulder-high, leafy rows are festooned with pendulous bunches of plump table grapes. "Taste these," he says, proffering a handful of sweet, succulent fruit. The shadow from his straw cowboy hat barely conceals the look of pride creasing his sun-browned face. "Better living through chemistry? Who says?"

Pavich is among the increasing number of commercial farmers who have shunned chemical pesticides. One alternative they've adopted is integrated pest management (IPM). Rather than attempt to eradicate all pests, IPM assumes that natural processes and natural enemies are essential to a healthy farm. Experts liken it to using a scalpel rather than a cleaver—by monitoring and sampling their fields more closely, farmers can employ highly focused suppression programs that take aim at specific pests using few or no chemicals.

Frank Zalom, statewide IPM director for the University of California's agriculture program, explains why farmers are seeking alternatives to chemicals: "First, government regulations have placed increasing restraints on the use of pesticides, forcing farmers to figure out ways to cut down their sprays. Second is the public outcry about pesticides. Farmers are reacting to consumer concerns."

Zalom says that 60 percent of California tomato growers now practice some form of IPM and that in the state's strawberry fields, the use of biological controls to manage pests is up nearly 70 percent in five years.

Biological control is one of the most promising tools in the IPM arsenal. Its premise is simple: Naturally occurring predators, parasites, and pathogens are used to fight pests, weeds, and disease. Strawberry growers in California, for instance, deploy predatory mites in their fields to control the pesky Pacific spider mite.

Using good bugs to eat bad bugs is hardly new. As early as 324 B.C., Chinese farmers placed nests of the predatory ant *Oecophila smaragdina* in citrus trees to dominate caterpillars and boring beetles. The growers even went to the trouble of linking trees with tiny bamboo bridges to make the patrolling ants' job easier. Today China is the world's leading practitioner of biological control, relying on beneficial insects to manage pests on more than 21 million acres of cropland.

Citrus trees were the target of the first successful attempt at biological control in the United States as well. In the 1880s a devastating insect known as cottonycushion scale attacked California's orange groves. The infestation was so thick in some regions that the trees looked as if

they were covered with snow. Relief finally came when entomologists discovered an Australian ladybug that was the scale's natural enemy. Other triumphs followed: "Good" bugs were used to protect sugarcane in Hawaii, citrus in Florida, walnuts in California, and 2 million acres of rangeland in the West.

Kicking the Habit

Despite these successes, biological control never caught on in the United States because of the invention of DDT 50 years ago and the ensuing national addiction to chemical pesticides. But as growing concern over air, water, and food quality has brought about stiffer pesticide regulation, biological control is helping many farmers kick their chemical dependencies.

Take grape grower Pavich, for ex-

ample. In the late 1970s his farm was hit by recurring waves of voracious grape leafhoppers and Pacific spider mites. "Just like I was taught to do in college," he recalls, "I called in a representative from a major chemical company. It almost ruined us."

The salesperson prescribed bombarding the farm with a lethal chemical. The result was instantaneous—nearly all the pests were killed—but those that survived succeeded in spawning a new generation resistant to the poison. A kind of superbug evolved, overrunning the farm in no time. Additional spraying of more potent pesticides just made matters worse.

Pavich found himself on a treadmill. He was being forced to pay more and more money for pesticides that were less and less effective. Out of desperation he looked for an alternative and began to read up on biological control. Deciding to give it a try, he drove to the nearby mountains, collected thousands of ladybugs, and released them in his fields. The gentle red beetle of nursery-rhyme fame is actually a bug-munching machine that begins feeding the moment it leaves its orange egg, consuming 40 aphids an hour. To augment his entomophagous army, Pavich enlisted Chinese praying mantises.

"It was trial and error at first," he says, "but we finally were able to restore the farm's natural balance." The success induced Pavich to try other benign pest-control techniques nearly forgotten in the chemical age: crop rotation, intercropping, soil flooding, improved tillage, and building up the soil with natural nutrients. Six years ago he made the complete transition to organic farming, banning chemicals from his entire operation.

Pavich says the benefits far outweigh the costs. "For one thing, we don't have to support a big chem bill anymore," he explains. Grape growers routinely spend up to $600 per acre for a chemical mix of fungicides, insecticides, and herbicides. "Now we put that money into other areas, like nutrients. The result is that our fruit tests 100 percent to 400 percent higher in nutrition." That is one of the reasons that Pavich can sell his grapes for 10 to 15 cents more per pound than nonorganic grapes.

Production hasn't suffered, either. The National Research Council, an agency sponsored in part by the National Academy of Sciences, issued a study in September 1989 reporting that farmers who apply few or no chemicals to crops usually get as great a yield as those who use pesticides and synthetic fertilizers. Pavich says his lands are even more fruitful than they used to be.

Farming without chemicals does have some drawbacks. It is more labor-intensive and requires a much higher level of field management. Benign alternatives don't act as quickly or as powerfully as chemical pesticides, nor do they last as long. But Pavich is able to save some of his additional labor costs by not having to monitor his field hands for exposure to pesticides.

Bugs as Business

The federal government has recently launched several new IPM research programs. One of the biggest is the Low Input Sustainable Agriculture program, or LISA. Approximately $5 million a year funds nearly 50 different projects.

Private research is also under way. Dozens of companies with an eye on capturing part of the $16-billion worldwide pesticide business are working to create chemical-free alternatives. They include commercial insectaries and biotechnology labs. Rincon-Vitova of Ventura, California, is already selling $1.5 million worth of beneficial insects a year. "Our *Trichogramma* bee attacks 250 species of moth and butterfly eggs," says company spokesperson Jake Blehm. "We sell bugs to everyone from Walt Disney World to the country's largest pecan ranch."

Biosys, a Palo Alto, California, firm, hopes to sell $50 million worth of nematodes five years from now. The company has hit upon a way of growing the ½-millimeter-long parasite by the hundreds of billions. The worms are dried but kept alive under carefully controlled conditions, allowing them to be packaged and stored for up to six months. They can be sprayed on gardens to kill white grubs and cut-worms.

Much of today's biopesticide development is focusing on a common soil bacterium called *Bacillus thuringiensis*, which has a seemingly unlimited supply of natural toxins. It is lethal to plant-gobbling pests, but, according to both the U.S. Department of Agriculture and the Environmental Protection Agency, it is safe for human and animal consumption. Several companies are modifying naturally occurring strains with gene-splicing techniques to beef up toxicity, prod them into producing more effective offspring, and make them tastier to bugs.

Classical biological-control experts such as the University of California's Donald Dahlsten are not convinced that genetic engineering is the way to pursue pests. "If you believe in evolution, you know that things adapt," he explains. "Put a genetically altered plant in the field, and in time the pests will adapt to it. Then what will we have? Biotechnology isn't going to solve our problems, just as DDT didn't."

Whether designer bugs prove more effective than those that occur naturally remains to be seen, but there's no denying that beneficial insects are a welcome alternative to chemical pesticides: There are now 68 suppliers of "good" bugs nationwide, sharing a $25-million world market. Farmers are discovering that biological diplomacy, rather than chemical warfare, makes for better agriculture.

From *Sierra*, January/February 1990. Copyright © 1990 Dwight Holing. Reprinted by permission.

EARTH CARE ACTION

Waiter, There's a Duck in My Soup

Cattle and pig farmers may soon find Muscovy ducks more common both on and off the dinner table. Such farmers have a problem with houseflies that invade farms in warm weather. They make life unpleasant, and they spread disease. The usual way to keep their level down is to spray them with chemicals, but this is losing its appeal because of the rise of two species: insecticide-resistant flies and environmentalists.

So when they read in a magazine about a farmer who used Muscovy ducks to control flies, Dr. Gordon Surgeoner and Barry Glofcheskie of the Department of Environmental Biology at the University of

Guelph in Ontario, Canada, set out to determine whether this odd idea is also a good one. Their results show that it is. In fact, ducks are better than flypaper.

Muscovy ducks come originally from South America and are thought to be the ancestors of all other domestic ducks. Muscovy ducks are omnivores, eating both plant and animal matter. They are particularly fond of insects.

Fast Eaters

In their first laboratory test, Dr. Surgeoner and Glofcheskie put an unfed five-week-old Muscovy in an 8-cubic-foot screened cage with 400 live flies. After an hour, the duck had eaten 326 of them. A subsequent experiment involved four ducks in separate cages filled with 100 live flies each. In each case, the ducks devoured over 90 percent of them within 30 minutes. Flypaper, flytraps, and bait cards took 15 to 86 hours to do the same job.

The experiments then moved outdoors. Pairs of two-year-old Muscovy ducks were put with the cows on several farms. When the ducks were around, there were 80 to 90 percent fewer flies. Videotapes of the ducks showed that they snapped at flies about twice a minute and enjoyed a healthy 70 percent kill ratio.

Then the experiments became more refined. The researchers discovered that female ducks seem to eat about 10 percent more flies than male ducks, and that ducks of any age between eight days and two years were fine for the job. The ducks adapt well to the barnyard scene, sticking close to the younger pigs and cows to which flies are particularly attracted—even snatching flies off their hides as they rested or slept. At one pig farm, the ducks huddled between sleeping piglets and were accepted by the sow. This camaraderie is a clear advantage over other fly-catching devices, which must be kept away from animals.

A Cheap Alternative

There are only two problems with the killer ducks. One is their tendency to peck at urethane insulation inside barns; the other is that they produce yet more manure. But their economic advantages are clear. The chemical control of flies costs $170 to $450 Canadian ($150 to $390 U.S.) a season for a 35-cow dairy; Muscovy ducklings cost $2 Canadian ($1.75 U.S) each, eat for free, and can be sold after the fly season for two to four times their original price. It is not yet known whether fly-fed Muscovies taste better or worse than their grain-fed cousins, but they weigh more and have excellent muscle tone. They certainly taste better than flypaper.

Dr. Surgeoner thinks good sanitation is still the best way to keep flies down, but as a second line of defense, ducks appear to be more effective than chemicals. Although using a few ducks will not eliminate the need to use insecticides, they can cut the amounts used. And ducks, at least, are biodegradable.

From *The Economist*, November 18, 1989. Reprinted by permission.

An iceberg off the Antarctic coast plays host to a number of Adelie penguins, one of the continent's wide variety of birds and other wildlife. (Greenpeace/Morgan)

ANTARCTICA

By Michael D. Lemonick, with Andrea Dorfman

From atop a windswept hill, the panoramic landscape looks eerily beautiful—and yet completely hostile to life. Even at the height of summer, the scene is one of frigid desolation. To the west lies a saltwater bay whose surface is frozen solid. Beyond the bay loom glittering glaciers and towering, rocky peaks. On the south and east rises a blinding white shelf of permanent ice, so thick that it grinds against the seabed far below. And to the north is a snow-covered volcano that continuously belches noxious fumes. This is McMurdo Station, at the bottom of the world, where winds can reach 320 kph (200 mph) and temperatures can plunge below −85°C. This is Antarctica, the white continent, the harshest, most forbidding land on earth.

But the view from the hilltop, overlooking McMurdo Sound on the eastern side of Antarctica, is deceiving. A closer look at the seemingly lifeless landscape and seascape reveals an amazing abundance of life. Like most of the coastal waters around the continent, McMurdo Sound is filled with plankton and fish, and its thick ice is perforated by the breathing holes of Weddell seals. Nearby Cape Royds is home to thousands of Adelie penguins, which hatch their eggs in the world's southernmost rookery. Skuas—sea gull-like scavenger birds—scout the breathing holes and the margins between sea ice and land, seeking seal carcasses and unguarded baby penguins to feast on. The ice itself is permeated with algae and bacteria.

News of environmental assaults has unleashed global concern about Antarctica's future. It is now clear that the continent's isolation no longer protects it from the impact of man.

There is another sort of life as well. All around Antarctica the coast is dotted with corrugated-metal buildings, oil-storage tanks, and garbage dumps—unmistakable signs of man. No fewer than 16 nations

93

have established permanent bases on the only continent that belongs to the whole world. They were set up mainly to conduct scientific research, but they have become magnets for boatloads of tourists, who come to gawk at the peaks and the penguins. Environmentalists fear that miners and oil drillers may not be far behind. Already the human invaders of Antarctica have created an awful mess in what was only recently the world's cleanest spot. Over the years, they have spilled oil into the seas, dumped untreated sewage off the coasts, burned garbage in open pits, and let huge piles of discarded machinery slowly rust on the frozen turf.

News of environmental assaults has unleashed a global wave of concern about Antarctica's future. "It is now clear that the continent's isolation no longer protects it from the impact of man," declares Bruce Manheim, a biologist with the Environmental Defense Fund. How best to protect Antarctica has been a topic of fierce debate in meetings from Washington to Wellington, New Zealand. Everyone agrees that the issue is of great importance and urgency. Despite the damage done so far, Antarctica is still largely pristine, the only wild continent left on earth. There scientists can study unique ecosystems and climatic disturbances that influence the weather patterns of the entire globe. The research being done on the frozen continent cannot be carried out anywhere else. "In Antarctica we still have the chance to protect nature in something close to its natural state and leave it as a legacy for future generations," says Jim Barnes, a founder of the Antarctic and Southern Ocean Coalition, an alliance of more than 200 environmental groups.

The focus of contention at the moment is the Convention on the Regulation of Antarctic Mineral Resource Activities— also called the Wellington Convention—an international agreement that would establish rules governing oil and mineral exploration and development in Antarctica. Proponents say the convention, painstakingly drafted during six years of negotiations, contains stringent environmental safeguards. But many environmentalists see the convention as the first step toward the dangerous exploitation of Antarctica's hidden store of minerals. They argue that the continent should be turned into a "world park" in which only scientific research and limited tourism would be permitted.

That position did not garner much support until last spring, when France and Australia, two countries with a major presence in Antarctica, suddenly announced that they backed the world-park idea and would not sign the Wellington Convention. In Washington, Sen. Albert Gore, Jr., of Tennessee, is leading a drive to get the United States to withdraw its support of the accord. Until the debate is resolved, there will be no agreed-upon strategy for protecting Antarctica from mineral exploration.

Trouble High in the Sky

Meanwhile, some of the harm already done will not easily be repaired and may have far-reaching impact. For many years, the industrial nations have been releasing chlorofluorocarbons (CFCs) into the atmosphere, not realizing that these chemicals were destroying the ozone layer, which shields the earth from harmful ultraviolet radiation. Because of the vagaries of air currents, ozone depletion has been most severe over Antarctica. It was the discovery in 1983 of an "ozone hole" over the continent that first alerted scientists to the immediacy of the CFC threat.

Since then, researchers have been monitoring the hole and looking for similar ozone destruction over populated areas. Scientists predict that thinning ozone, and the resulting increase in ultraviolet radiation, will cause damage to plants and animals, as well as skin cancers and cataracts in humans. To keep a bad situation from getting worse, nations are working on an international agreement designed to phase out production of CFCs by the year 2000.

In the meantime, researchers have been carefully studying the effects of ozone depletion on antarctic life. Marine ecologist Sayed El-Sayed of Texas A&M University discovered two years ago [1987] at Palmer Station, a U.S. base on the Antarctic Peninsula, that high levels of ultraviolet radiation damage the chlorophyll pigment vital for photosynthesis in phytoplankton, slowing the marine plants' growth rate by as much as 30 percent. That in turn could threaten krill, shrimplike creatures that feed on phytoplankton and are a key link in Antarctica's food chain. Says El-Sayed: "Fish, whales, penguins and winged birds all depend very heavily on krill. If anything happened to the krill population, the whole system would collapse."

The fragility of life in the antarctic climate was dramatically underscored last January [1989], when the *Bahia Paraiso,* an Argentine supply and tourist ship, ran aground off Palmer Station, spilling more than 643,450 liters (170,000 gallons) of jet and diesel fuel. The accident killed countless krill and hundreds of newly hatched skua and penguin chicks. Some 25 years of continuous animal population studies run by scientists at Palmer may have been ruined. Just weeks after the *Bahia* incident, the Peruvian research and supply ship *Humboldt* was blown by gale-force winds onto rocks near King George Island, producing an oil slick more than half a mile long.

Such disasters are shocking and unsettling to the hundreds of scientists in Antarctica, who had hoped the continent would remain their unspoiled natural laboratory. But they too bear much of the responsibility for the pollution that has soiled the area. Just three months ago, McMurdo Station, a U.S. base operated by the National Science Foundation, reported that 196,820 liters (52,000 gallons) of fuel had leaked from a rubber storage "bladder" onto the ice shelf. Over the past year or two, many bases have launched extensive cleanup campaigns, but scientists have yet to find the right balance between studying the Antarctic and preserving it.

The World's Largest Laboratory

No one disputes the importance of the research. The continent has a major—though not completely understood—influence on the world's weather. As Antarctica's white ice sheet reflects the sun's heat back into space, an overlying mass of air is kept frigid. This air rushes out to the sea, where the earth's rotation turns it into the roaring forties and the furious fifties—old sailors' terms for the fierce winds that dominate the oceans between 40° and 60° south latitudes. If scientists can figure out just how these winds affect the global flow of air, then it will be easier to understand and predict the planet's weather.

Antarctica also provides the best-preserved fossil record of a fascinating chapter in the earth's history. Some 200 million years ago, during the Jurassic period, Antarctica formed the core of the ancient supercontinent now known as Gondwanaland. The name comes from Gondwana, a region in India where geological evidence of the supercontinent's existence was found. At the time of the supercontinent, Antarctica was nestled in the temperate latitudes and was almost tropical. It was covered by forest and filled with reptiles, primitive mammals, and birds. But by 160 million years ago, the supercontinent had begun to break up. While most of the pieces, including South America, Africa, India, and Australia, stayed in warm regions, Antarctica drifted to the South Pole.

Thus was created the world's largest stretch of inhospitable land. Precipitation is so sparse over Antarctica's 14 million square kilometers (5.4 million square miles) that it is classified as one of the world's driest deserts. Because most of the small amount of snow never melts and has accumulated for centuries, 98 percent of Antarctica is permanently covered by a sheet of ice that has an average thickness of 2,155 meters (7,090 feet). That accounts for 90 percent of the world's ice and 68 percent of its fresh water. Although the sun shines continuously in the summer months, the rays hit the land at too sharp an angle to melt the ice. At the South Pole, the average temperature is −49°C (−56.2°F) and the record high is 13.6°C (7.5°F). During the per-

petual darkness of winter, the temperature falls to almost inconceivable levels. The lowest ever recorded was in 1983 at the Soviet Union's Vostok Base: −89.2°C (−128.6°F).

Around the edges, though, Antarctica is more than just an icebox. On the Antarctic Peninsula, which reaches like a finger to within 965 kilometers (600 miles) of South America, the temperature has risen as high as 15°C (59°F). The peninsula is home to the continent's only two species of flowering land plants, a grass and a pearlwort. Off the coast is one of the world's most productive marine ecosystems. Antarctica supports 35 species of penguins and other birds, 6 varieties of seals, 12 kinds of whale, and nearly 200 types of fish.

Explorers and Exploiters

It was the bountiful sea life that initially drew large numbers of men to the southern continent. When James Cook first circled Antarctica between 1772 and 1775, he saw hordes of seals on the surrounding islands, and during the next century the continent became a hunter's paradise. By the early 1900s, elephant and fur seals were nearly extinct. And after 1904, more than a million blue, minke, and fin whales were harpooned in antarctic waters.

Along with the exploiters came explorers, searching for nothing more than scientific knowledge and personal and national glory. In 1841, Britain's James Clark Ross became the first man to find his way through the sea ice and reach the mainland. The ultimate goal for the adventurers—the South Pole—was not reached until seven decades later, during the dramatic and ultimately tragic race between British explorer Robert Scott and Norway's Roald Amundsen. Relying on dogsleds, which proved to be more dependable than the breakdown-prone mechanical sleds used by Scott, Amundsen's party arrived triumphantly at the pole on December 14, 1911. When Scott got there a month later, he was devastated to find a Norwegian flag flying and notes from Amundsen. Things got even worse on the way back. Only 18 kilometers (11 miles) from a supply depot, Scott and two companions were stopped by a blizzard, their fuel and food nearly gone. Scott's diary entries end this way: "We shall stick it out to the end, but we are getting weaker, of course, and the end cannot be far. It seems a pity, I do not think I can write more. . . . For God's sake look out for our people."

There could easily have been major territorial conflict, but scientific cooperation intervened.

Airplanes made antarctic travel much less perilous. In 1929 Richard Byrd, an American, became the first person to fly to the South Pole, a 16-hour round trip from the west coast of Antarctica. And in the 1930s, German aviators claimed part of the continent for the Third Reich by dropping hundreds of stakes emblazoned with swastikas.

The postwar German government did not press the Nazis' claim, but seven other nations with histories of antarctic explora-tion—Argentina, Chile, France, New Zealand, Britain, Norway and Australia—maintained that parts of the continent belonged to them. Some of the claims overlapped: Chile, Britain, and Argentina, for example, all declared their ownership of the Antarctic Peninsula. The United States, while making no claims, refused to recognize those of other nations and organized numerous expeditions, including the largest in antarctic history. Mounted in 1946 and called Operation Highjump, it was a naval exercise involving 13 ships, 50 helicopters, and nearly 5,000 service members. Its unstated purpose: to make sure the United States could legitimately stake its own claim should it ever want to do so.

There could easily have been major territorial conflict, but scientific cooperation intervened. It took the form of the International Geophysical Year (IGY), actually 18 months long, which was scheduled to take advantage of the peak of sunspot activity predicted for 1957 and 1958. Sixty-seven countries joined in this exhaustive study of the interactions between the sun and earth. Much of the research went on in Antarctica, where Argentina, Australia, Belgium, Chile, France, Britain, Japan, New Zealand, Norway, South Africa, the United States, and the Soviet Union established bases.

The antarctic component of the IGY worked so well that after the project ended, President Dwight Eisenhower invited the 11 other nations that had built bases to join the United States in an agreement that would govern all activities on and around the frozen continent. The resulting Antarctic Treaty, ratified in 1961, forbids military activity, bans nuclear explosions and radioactive waste disposal, and mandates international cooperation and freedom of scientific inquiry. Moreover, those participating countries that claimed chunks of Antarctica as their own agreed not to press those claims while the treaty remained in force. Over the years, 13 other countries have become voting members of the treaty system, and the original document has been supplemented by agreements governing topics as diverse as waste management and the protection of native mammals and birds.

The treaty did not eliminate the jockeying for position. The United States and the Soviet Union have deliberately placed bases in areas claimed by others, and countries have tried to solidify their stakes by setting up post offices and sending children to school in Antarctica. Argentina flew a pregnant woman to its Marambio base so that she could give birth to the first native of Antarctica. But no nation has overtly asserted sovereignty since the 1950s. Even during the Falklands war, Britain and Argentina, together with other nations, sat down to discuss Antarctic Treaty issues.

Some Call It Home

Amid an atmosphere of international partnership, research has flourished. In the past few weeks alone, Antarctica's scientists have carried out dozens of unique experiments. In the McMurdo Sound area, a group of geologists camped out in the bitter cold of the Royal Society mountains, looking for evidence of the ebbing and flowing of glaciers in Antarctica's past, and biologists drew 50-kilogram (110-pound) fish from ice holes to study the unique organic antifreeze that keeps these sea dwell-

ers alive. Volcanologists braved the knifelike winds and choking fumes atop Mount Erebus to learn what kinds of gases and particles Antarctica's largest volcano emits. At Williams Field, a runway on the Ross Ice Shelf, a multidisciplinary team prepared to launch a huge helium balloon. Its purpose: to follow circumpolar winds around the entire continent, gathering data on cosmic rays and solar flares and testing the behavior of high-density computer chips in the intense radiation of the upper atmosphere. And deep in the interior, glaciologists at the Soviets' Vostok Base dug out ice samples that carry clues to the planet's atmosphere in layers laid down in the polar ice cap tens of thousands of years ago.

At the South Pole, meanwhile, astrophysicists were taking advantage of a heat wave—the temperature had soared to −23°C (−10°F)—to set up detectors that would peer at the faint microwave radiation left over from the Big Bang explosion, which theoretically started the universe. In the high altitudes atop the pole's ice cap, the detectors are well above the densest, murkiest layers of atmosphere and can peer through some of the driest, clearest air on earth to help determine whether the original Big Bang was unique or was followed by smaller ones. A few hundred yards away, close to the enormous geodesic dome that covers the thickly insulated buildings of the U.S.'s Amundsen-Scott South Pole Station, atmospheric scientists measured traces of pollutants released around the globe. The pole is so remote from civilization that there, better than anywhere else, scientists can accurately assess just how far-reaching are the effects of pollution.

The researchers who seek such knowledge are adventurous souls who know better than most the meaning of the term hardship post. Counting construction workers, maintenance crews, and other support staff, Antarctica's population is only 4,000 or so, even in midsummer. The scientists and other residents tend to be in their twenties and thirties—vigorous enough to endure the world's coldest workplace. A carpenter's helper recalls toiling one time at –40°C (–40°F) in an unheated building. She had on so many layers of clothing that it took most of her energy just to move, she says. As for the scientists, common sense sometimes gives way to a sense of mission. Researchers handling delicate experiments have been known to work without gloves in subfreezing temperatures until their hands were numb.

Just as daunting as the cold are the loneliness and isolation in a land where phone lines are rare, mail is erratic, and penguins vastly outnumber people. Thousands of miles from friends and families, the residents of Antarctica are often confined to small areas around their bases. At many stations, living quarters are built underground so that they are protected from the wind. When storms force workers to stay indoors for days at a time, it amounts to their being trapped in a bunker.

But the bases try to make antarctic life as enjoyable as possible. At McMurdo Station, the continent's largest town, the 1,100 or more summer residents can hang out at the four Navy bars, use a two-lane bowling alley, take aerobics classes at the gym, and borrow videotapes from a library. Recent social events included a chili-cooking contest and a amateur comedy night. Even at the South Pole Station, home to no more

than 90 hardy workers, there is an exercise room, a sauna, a poolroom, and a library equipped with wide-screen TV and a VCR.

Along about February the annual exodus begins in earnest. Once the cold season takes hold, planes stop making regular flights to inland stations, and the ice layer spreads out to sea, making access by ship nearly impossible. Only a few hundred residents stay through the winter.

The number of people who have gone to Antarctica is smaller than the attendance at the 1990 Rose Bowl game, but those few have had a disproportionately large impact. Because plants and animals, along with human outposts, are largely confined to the 2 percent of Antarctica that is ice-free for part of the year, the world's most sparsely populated continent is, paradoxically, overcrowded. The Antarctic Peninsula is particularly in demand, with 13 stations; King George Island, one of the South Shetland Islands, is home to an additional 8 stations. Planes, helicopters, snowmobiles, trucks, and bulldozers are in constant operation throughout the summer. Nearly every base has its own helipad, landing strip, harbor, and waste dump.

Nasty Habits Exposed

The inhabitants of these bases have been notoriously careless, often discarding trash in ways that would be illegal at home. But their actions went largely unnoticed until January 1987, when Greenpeace became the first nongovernment organization to establish a permanent antarctic base, located at Cape Evans, some 24 kilometers (15 miles) north of McMurdo Station. The group has mounted annual inspection tours of dozens of bases. It was Green-peace that publicized McMurdo's contin-ued dumping of untreated sewage into the sea and burning of trash in an open-air pit. The waters right off the station are report-edly more polluted with substances such as heavy metals and PCBs than any similar stretch of water in the United States. Greenpeace has also documented reckless dumping and burning at Soviet, Uru-guayan, Argentine, Chilean, and Chinese bases. And an airstrip under construction at France's Dumont d'Urville base has al-ready leveled part of an Adelie penguin rookery.

While scientists try to clean up their act, tourists are posing an increasing threat to Antarctica's delicate ecosystems. Chilean planes began flying in visitors in 1956, and luxury cruises started a decade later.

The charges have some validity, says Erick Chiang, senior U.S. representative in Antarctica, but they are exaggerated. "Our behavior in the past was disgraceful—by today's standards," he admits. "But we are doing much better. We're installing a pri-mary waste-treatment facility at McMurdo this season. We've begun recycling. Yes, we lost 50,000 gallons of fuel recently, but we've recovered more than half of it." Last month McMurdo residents went patrolling for loose trash.

Chiang contends that despite past sins, the local ecology has not suffered very much. Some scientists agree. Says Corne-lius Sullivan of the University of Southern California, who studies the algae that live in and under McMurdo Sound ice: "A few places are filthy. But most of the water is still absolutely pristine." Nonetheless, the National Science Foundation could do much better. One thing that will help: About $10 million was added to the agen-cy's budget for 1990, bringing it to $152 mil-lion, and much of the new money will go toward protecting the environment.

While scientists try to clean up their act, tourists are posing an increasing threat to Antarctica's delicate ecosystems. Chilean planes began flying visitors to the penin-sula in 1956, and luxury cruises started a decade later. Although commercial flights stopped after an Air New Zealand DC–10 crashed into Mount Erebus in 1979, killing all 257 aboard, ship travel has thrived. About 3,500 people, mostly Americans, paid $5,000 to $16,000 to sail over from South America last year. They generally stayed in Antarctica four or five days. Most boats carry naturalists or other experts, who give lectures, and groups often visit scientific stations. So many boats cruise along the peninsula between November and March that it has been dubbed the "Antarctic Riviera." Chile has opened a ho-tel near its base. Antarctic activities include hiking, mountain climbing, dogsledding, camping and skiing. A few show-offs have even water-skied on the cold waters.

The most intrusive visitors are those who tramp through penguin rookeries and other wildlife habitats. Going anywhere near certain kinds of seabirds can frighten them enough to disrupt feeding patterns and reproductive behavior. Though warned not to litter, some tourists leave behind film wrappers, water bottles, and cigarette butts. And yes, Antarctica has graffiti—on the rocks of Elephant Island.

Responsible tour operators have come up with a code of conduct that forbids visitors to harass animals, enter research stations unless invited, and take souvenirs. Preservationists, like the Environmental Defense Fund's Manheim, argue in addition for strict limits on the size and frequency of tours and for civil and criminal penalties for operators who do not comply with the rules.

The Antarctic Treaty nations may discuss tourism when they meet later this year [1990], but they are more likely to be preoccupied with the growing debate over the future of oil and mineral development. Concern first arose after the 1973 oil crisis, when it became clear that there might someday be pressure to drill for petroleum, even in the harsh antarctic environment. Eventually, the treaty nations decided it was best to have rules in effect before that happened. The result was the Wellington Convention, agreed to by representatives of 20 treaty nations in New Zealand's capital in June 1988. The document essentially forbids any mineral exploration or development without agreement by all treaty participants. But most environmentalists are disturbed by any accord that recognizes even the possibility of oil drilling. Naturalist Jacques-Yves Cousteau has called the Wellington Convention "nothing more than a holdup on a planetary scale."

There is no certainty that commercially valuable deposits of minerals exist. Surface rocks contain traces of iron, titanium, low-grade gold, tin, molybdenum, coal, copper, and zinc. Gaseous hydrocarbons, sometimes associated with oil, have been found in bottom samples taken from the Ross Sea. But in most cases, says geologist Robert Rutford, president of the University of Texas at Dallas, who did research in Antarctica for more than 20 years, "minerals are less than 1 percent of the total rock sample analyzed." Moreover, the vicious antarctic climate would make exploration dangerous and expensive.

Still, say the Wellington Convention's opponents, some countries might be tempted anyway. Contends Barnes of the Antarctic and Southern Ocean Coalition: "Some nations are awash in cash and technology and have no domestic oil supply. I think Japan would be down there as soon as the continent was opened up." Opponents of drilling point out that the Antarctic Treaty has not always been scrupulously adhered to, especially when it comes to fishing limits and environmental protection. They argue that the Wellington Convention could also be skirted.

Support for a World Park

Such arguments are behind the surge in support for a world park. The proposal by Australia and France last October that the continent be declared a "wilderness reserve" under the eye of an antarctic environmental protection agency—essentially the world-park scheme by a different name—was hailed by environmentalists as a big victory. The United States, still officially committed to the Wellington agreement, did not go along with the new initiative. But some Administration officials are said to be opposed to the minerals convention, and Sen. Gore claims he has the votes to prevent its ratification in the Senate. Observes Gore: "The whole theory of protecting Antarctica with mining that is carefully circumscribed by safety procedures is the approach that failed in Alaska's Prince William Sound. We shouldn't make the same mistake again."

Nonratification by either France or Australia would automatically kill the Wellington Convention. But that does not guarantee that the world-park concept, as good as it would be for Antarctica's environment, would replace the defeated agreement. Some Antarctic Treaty nations oppose a permanent ban on mineral development—notably Britain, which has the same veto powers as France and Australia. That raises the possibility that the world will be left with no agreement at all on the minerals question, not even the informal moratorium on exploration and mining adopted in 1977 until a convention could be ratified. Antarctica might thus be opened to wholly unregulated mining.

That is a frightening prospect, so alarming that the nations subscribing to the Antarctic Treaty cannot afford to let it happen. The Wellington Convention may not be perfect, but it should be ratified. Far from a license to exploit, it would serve as a major roadblock to development and could be strengthened by further conventions specifying more stringent protection—even by the creation of the same environmental watchdog agency suggested by world-park proponents. The real problem with the Antarctic Treaty system is that the rules are not always strictly enforced, and there is no reason to think that nations would pay any more attention to the provision of a world-park system than they do to existing regulations.

In the end, the only way to save Antarctica is to convince the countries operating there—and those that join them in the future—that it is not worth fouling the only relatively untouched continent left on earth to gain a few extra barrels of oil. The environmental activists have done much to make that point, and governments seem to be listening. This may be the place where mankind finally learns to live in harmony with nature. If so, the forbidding vistas of Antarctica may be just as full of life a century from now as they were when humans first set foot on that continent less than 200 years ago.

Editor's Note: The National Wildlife Federation and many national conservation organizations oppose ratification of the Convention on the Regulation of Antarctic Mineral Resource Activities (CRAMRA) because its process for approving mining and oil and gas development indicates that these activities will be permitted in the future. The conservation organizations prefer a comprehensive protection agreement as proposed by Australia and France, which would make the antarctic continent a "wilderness park" not subject to exploitation and development.

The Arctic National Wildlife Refuge

By Brian K. Vincent

Along the northeast slope of Alaska there lies a land of majestic mountains, braided rivers, and endless sky. Unspoiled and untamed, it is a place where the dance of life moves freely; where caribou and wolves carry on their primordial drama, the mighty grizzly bear lumbers across the open tundra, and muskoxen—relics of the Ice Age—brave the howling blizzards of the long winter. This is the Arctic National Wildlife Refuge.

In the late 1950s, Supreme Court Justice William O. Douglas visited Alaska's northeast slope at the invitation of renowned wildlife biologist Olaus Murie. So struck was he that he persuaded President Dwight Eisenhower to set aside 8 million acres as the Arctic Wildlife Range. With the passage of the Alaska National Interest Lands Conservation Act in 1980, Congress expanded the range to 19 million acres and renamed it the Arctic National Wildlife Refuge. Only the 1.5-million-acre coastal plain, a narrow strip of land cradled between the Brooks Range and Beaufort Sea, was excluded from wilderness protection, pending study of the area's potential petroleum reserves.

The coastal plain represents the last undeveloped stretch along the northern slope of Alaska and is home to more than 160 animal species. Snowy owls, golden eagles, and arctic foxes make the coastal plain their home year-round. Migratory birds, including swans and snow geese, use the plain as a resting area before heading to points far away. Perhaps the greatest wildlife spectacle is the annual migration of 185,000 caribou from Canada to calve on the coastal plain. This is often compared to the migration of wildebeests across the African plains, and the area has earned the distinction of being called "the American Serengeti." A U.S. Department of Interior report calls the plain the center of wildlife activity in the refuge.

Despite this conclusion, the report recommended opening up the entire coastal plain to oil and gas leasing. But it also predicted oil development could result in a major population decline of the caribou herd, muskoxen, and lesser snow geese, as well as "substantially decreased habitat value for denning polar bears." Increased access could result in hunting and trapping of the refuge's remaining wolves, which, along with grizzly bears, have disappeared from the nearby Prudhoe Bay oilfield. The Department of Interior concedes there is only a one in five chance of finding oil at all, much less in commercially large quantities.

103

Drilling Advocated

But oil companies contend there could be significant amounts of oil under the coastal plain, and that without increased domestic oil production, the United States will again be subject to the whims of OPEC. American Petroleum Institute president Charles DiBona testified before a joint hearing by the Department of Energy and the Department of Defense: "The nation's energy security is threatened by rapid growth in imports, now at essentially 50 percent of use. Under present policies, two-thirds or more of our supplies could be imported within a decade. In fact, growing moratoria on offshore drilling and lack of access to the refuge will accelerate the trend."

Environmentalists argue that the solution to meeting our nation's long-term energy needs lies in conservation measures, such as increased auto efficiency standards, weatherization, and alternative fuels. In addition, they say, the burning of nonrenewable fossil fuels contributes to global warming, necessitating the development of alternatives. "If one goal of our energy policy is to develop alternatives to our continued national addiction to finite world reserves of oil," testified Brooks Yeager of the National Audubon Society at the joint hearing, "let's also agree that it makes no sense to start that process by shooting up one last time, and in the process destroy the wilderness character of one of the world's last great arctic ecosystems."

Even with a well-organized lobbying effort by the Alaska Coalition, a group of environmental organizations working to protect the refuge, the Senate Energy and Natural Resources Committee passed a bill sponsored by Sen. J. Bennett Johnston of Arizona last March [1989] that would allow development of the coastal plain. Sen. William Roth of Delaware and Rep. Morris Udall of Arizona have introduced legislation that would preserve the entire coastal plain.

Industry had hoped to pass development legislation in 1989. But that was before the *Exxon Valdez* tragedy. The 11-million-gallon spill illustrated what many environmentalists had warned prior to the construction of the Trans-Alaska Pipeline: Oil and pristine environments don't mix. It also raised serious doubts about industry's ability to clean up spills and disturbing revelations about industry practices.

President Bush still maintains there is no connection between the spill and development in the refuge, even though arctic oil would be pumped down the pipeline to the Valdez terminal and transported by tanker. Former Alaska Coalition Chairman Tim Mahoney disagrees: "The *Exxon Valdez* is connected to the Arctic Refuge by the pipeline, the politics, and the promises. It is the same crude oil product, the same powerful producers, and the same daily pollution of the wilderness. It is the same promises and the same seductive song—instant national security, no environmental problem."

Liability Laws

After the spill, Congress shifted its attention from the Arctic to oil-spill liability and compensation legislation. Until the *Valdez* disaster, only a patchwork of laws was in place. The tragedy highlighted the need for a comprehensive federal cleanup policy that would ensure quick, effective cleanup of spills, and, more importantly, would prevent them.

Last November [1989], the House passed a bill that holds companies liable for spills and established a $1 billion industry-financed fund to pay for cleanup. The bill passed by a 357–5 vote. A House/Senate conference was scheduled to work out differences between their two versions of the bill. One controversial difference is the House requirement that all new tankers be built with double hulls and bottoms. The Senate has rejected the double-hull mandate, voting instead to study the issue further.

While efforts to drill have been derailed, the coastal plain is not yet safe. Oil-spill liability legislation will become law next year, but whether it is the solution to the problems made apparent by the *Valdez* is questionable. Industry spends $25,000 to $50,000 per week on public relations, calling Prudhoe Bay an "environmental success story" despite the fact that thousands of spills have been reported there. Next fall

[1990], the Bush Administration will unveil its national energy plan, of which development of the Arctic is sure to be a part.

No matter how carefully undertaken, oil development is a large-scale industrial activity. It requires huge quantities of gravel for roads, drill pads, and airports, and equally huge quantities of scarce fresh water from coastal plain rivers. Pollution from heavy machinery operating around the clock will have an impact on the fragile tundra far beyond the actual boundaries of the facilities.

The essence of the Arctic National Wildlife Refuge is wildlife in its wild setting. Environmentalists maintain that only by designating the entire Arctic National Wildlife Refuge as wilderness can this American Serengeti be spared the fate now facing Prince William Sound.

From *E—the Environmental Magazine* (1–800–825–0061), March/April 1990. Reprinted by permission.

 EARTH CARE ACTION

The End of Antarctica?

By Robert Hennelly

Antarctica, the only continent without a border, is crucial to our planetary survival, yet it is on the verge of becoming a "wild West" mining territory with no law

but the 200-mile-an-hour winds that blow across its icy surface.

Antarctica holds 90 percent of the earth's ice and is the major cold source that

Jacques Cousteau and his "crew" of six children, one representing each continent, survey the last frontier. (Photo courtesy of the Cousteau Society)

mitigates the heat of the sun as it beats through our tattered ozone.

Now Jacques Cousteau and his Cousteau Society are heading up a global campaign to block the adoption of the Wellington Convention Accords drafted in New Zealand in 1988. The Accords would open our last frontier for the mining of uranium, heavy metals, and coal, and for the drilling of oil.

"In July of 1988, I saw a small news item about the opening of Antarctica to mining. This is enormous news, and it was just a few lines. It seemed to me there had been a coverup," Cousteau said at a press conference.

Antarctica's 5.4 million square miles are the essential coolant element of the earth's thermodynamic machine. The average temperature there is –49°C (–56.2°F).

The Cousteau Society notes that "the antarctic glacial cap reflects up to 80 percent of the solar radiation it receives, thereby helping maintain low temperatures in the area. This capacity for 'manufacturing' cold in the heart of a dynamic system allows a regulation of the earth's mean temperature."

Cousteau contends that any further degradation of the continent would have grave global environmental impacts and further accelerate the greenhouse effect.

"This would be total folly. Any change in the color of the surface, even dust, and the ice would melt and cause the sea to rise 2 or 3 meters," Cousteau said. "They say you can prospect with small charges, or if you mine, you have to be clean. Who is going to implement these conditions? There is no authority to do it."

In 1959, the International Antarctic Treaty reserved the icy continent for only peaceful and scientific purposes. Thirty-nine countries have since signed the treaty. In June of 1988, these signatory nations adopted the first Convention on the Regulation of Antarctic Mineral Resource Activities at Wellington, New Zealand, opening the door to exploitation of its natural resources.

Treaty Is Not Enough

According to the Cousteau Society, "The negotiators of the Antarctic Treaty had not envisioned the insatiable appetite of modern man for energy and mineral resources, thus they did not think it necessary to regulate specifically this type of activity, thereby leaving a 'legal void.' This legal void, as it were, protected the Antarctic effectively for nearly 30 years, since no one could risk prospecting and exploring without being assured of a license to develop, which only international regulation can guarantee."

Cousteau wants nations to reject the accords, which he feels will lead to the destruction of the continent, thereby disrupting the planet's atmospheric equilibrium.

The prospect of drilling for oil on and near Antarctica deeply troubles him. "What would happen if *Valdez* had happened in Antarctica? In such places as cold and arctic, things remain, and oil spills will

do the same. The disaster will stay almost forever."

To underscore his point as to the lack of regulation and environmental oversight on the continent, Cousteau tells the story of the *Bahia Paraiso*, a supply ship that sank a year ago [1989] near the American scientific base at the South Pole. It sits in the same place to this day, leaking its 50,000 to 60,000 gallons of diesel fuel into the same pristine arctic waters. Cousteau says that, as a result, 2,000 birds died and years of scientific study were destroyed.

When asked if anyone had taken responsibility for the ship or if there had been any regulatory oversight of the cleanup, Cousteau responded, "As of today, no. Nothing. The Argentinians had a project, but a year has passed and nothing has been done."

As to whether anyone would be liable for the environmental damage and have to pay fines or penalties, he said, "No, if you want go there, just go. If you want to wreck the Antarctic, go ahead and wreck it. Nobody owns the Antarctic, and theoretically nobody can complain. So I ask, when there is a mining disaster, who will appeal? It is a very disturbing situation."

Presently, the last frontier is also having its waters overfished, as 100 Russian fishing boats and smaller flotillas from Chile and Japan go after the krill, a tiny shrimplike crustacean that is a major food source for penguins, seabirds, seals, whales, and fish.

Cousteau recently traveled to Antarctica with six children, one representing each continent. "The presence of these children in Antarctica is indispensable as a symbol. Their flag suggests a child embracing Antarctica," he said.

Cousteau has gotten 1.5 million signatures in France and 250,000 in the United States to protest mineral exploitation of Antarctica. Mining interests in Norway, the United States, and the United Kingdom are the most persistent in trying to open up our last frontier.

Do we yet have the character or the foresight to meet the challenge of preserving this vital element of our planet? Can we transcend territorial boundaries and greedy impulses to reach a protective international consensus on the fate of Antarctica?

Antarctica can provide a model for global environmental cooperation. Its exploitation can be the source of the planet's accelerated ecological demise.

 EARTH CARE ACTION

Antarctic Litterbugs Swatted

By David Clark Scott

Antarctic litterbugs beware! Tomorrow [January 10, 1990] the Greenpeace ship *Gondwana* embarks on the second leg of an Antarctic environmental policing and publicity tour.

The southern hemisphere summer draws about 4,000 scientists each year from more than 30 countries to Antarctica to research everything from the ozone hole to marine life. When they descend on the polar continent, their waste isn't always disposed of in accordance with internationally recognized Antarctic Treaty guidelines.

In October and November [1989], *Gondwana* crew members inspected 23 scientific stations belonging to 13 countries along the Antarctic peninsula. Since the last Greenpeace inspection two years ago, "a lot of bases have shown improvement," says Dr. Maj De Poorter, the campaign coordinator.

Greenpeace praises the changes in disposal practices of bases run by Argentina (Jubany base), Brazil, China, Poland, South Korea, and Uruguay. But low marks are given to the Soviet Union's Bellingshausen base and Argentina's Esperanza and Marambio bases.

Environmental activists remove fuel from rotting drums left behind by inhabitants of a British government research base on Antarctica's Deception Island. (Greenpeace/Morgan)

"Marambio was appalling," Dr. De Poorter says. The Greenpeace report states: "No effort has been made to follow the Antarctic Treaty's Waste Disposal Code. Gullies are strewn with waste matter literally thrown from buildings or pushed over the cliffs for 20 years. At present, there are no plans and no budget to start a cleanup."

The Soviet Union supply ship *Akademik Fedorov* was presented with 40 bags of food wrappers, plastic, glass, and old radio equipment collected by a Greenpeace crew from around the Bellingshausen camp. The Soviets were given credit for removing 200 barrels of oil and oil waste products but criticized for burning plastic, which is against the disposal code.

Britain drew rebukes for burning waste and for killing seals to feed the 29 dogs "kept mainly for recreational purposes" at Rothera Station. The United States Palmer Station was commended for "no signs of open burning anymore and no rubbish" in the area.

Adherence to the waste guidelines is especially critical, says Dr. De Poorter, because of the fragility of the environment.

The January/February expedition will resupply the four-person Greenpeace station at Cape Evans, then inspect McMurdo, the largest U.S. base, Italy's base at Terra Nova Bay, and the airstrip under construction at France's base at Dumont d'Urville.

During the Great Louisiana Toxics March of 1988, residents strode past petrochemical plants chanting, "Clean it up or shut it down!"—a sentiment shared by millions across the country. (Kittner)

IT'S ENOUGH TO MAKE YOU SICK

By Susan Q. Stranahan

April 1970. All across the nation, many Americans were discovering that pollution in its various forms was not merely an annoyance but a serious threat to their health, and they were mobilizing to do something about it.

That month, officials in the United States and Canada banned fishing on Lake Saint Clair and imposed a partial fishing ban on Lake Erie. For nearly three decades after World War II, industries in both countries had dumped millions of pounds of mercury and other poisons into the Great Lakes watershed, even though mercury's toxic properties were already well known. Indeed, a century earlier, Lewis Carroll made the Mad Hatter in *Alice's Adventures in Wonderland* a victim of mercury poisoning. Yet no one sought a solution until large numbers of fish caught in Lake Saint Clair showed mercury levels 14 times the amount deemed safe for human consumption.

Twenty years ago in April, federal officials also announced a ban on the home use of the herbicide 2,4,5-T and halted its use as a defoliant in Vietnam. The action came in response to reports of miscarriages, birth defects, respiratory difficulties, and neurological disorders among people living near areas where aerial spraying of the chemical was commonplace.

A few days after that announcement, Edward Simmons was manning a broom near the South Street Seaport Museum in Manhattan, helping to spruce up the city for something called Earth Day. "Last year I read the papers and figured I didn't have very long to live," Simmons, then 13, told a reporter at the time. "But now I'm beginning to think maybe I won't die so soon, because more people are seeing what's wrong—and something's got to happen."

Something did happen. The next two decades produced a tidal wave of new laws, scientific research, and technological advances, public awareness and activism—all geared toward making the United States a cleaner and healthier place to live. But the same period has also brought discoveries of environmental problems no one dreamed of in April 1970, dilemmas reflected in the

111

host of recent additions to the average American's vocabulary: acid rain, Superfund, carcinogens, mutagens, radon, dioxin, global warming; in place names that need no further identification: Times Beach, Love Canal, Bhopal; and in a list of new products expressed as ominous-sounding initials: PCBs, DDT, EDB, TCE.

---

*M**ost Americans today come in daily contact with a variety of chemicals suspected of posing health risks.*

To the exuberant celebrants of 1970, cleaning up the nation's pollution was achievable. Today, most environmental pragmatists agree that the less complex problems have been solved, while those that remain will be extraordinarily difficult to bring under control.

Without doubt, the United States has made great strides in protecting its population in the past 20 years. "You only have to go to central Europe or Latin America to see what America would be like if we hadn't imposed any [environmental] controls," says Ellen K. Silbergeld, a specialist in public health and biomedical research at the Environmental Defense Fund. But with each step, it becomes more obvious that we have only begun to understand the true extent of the toll pollution is taking on our lives. Most Americans today come in daily contact with a variety of chemicals suspected of posing health risks. Still, scientists cannot say with certainty at what levels exposure to most of them cause cancer,

neurological disorders, or birth defects— or whether they cause them at all.

"We're sort of in the cave-painting era in terms of what we know [about environmental health risks]," says Marvin Schneiderman, former associate director of the National Cancer Institute. Of the 48,000 or more chemicals registered with the Environmental Protection Agency (EPA), almost no health research has been done on 38,000 of them. About 1,000 have been tested to measure acute risks, while only half that many have been studied for their cancer-causing or reproductive effects or for their ability to cause genetic change.

At the same time, however, scientists have become more sophisticated detectives, capable of predicting which types of chemicals are most likely to threaten our health. "We also have gotten better at estimating the potency at which they cause problems," explains Ila Cote, who is associate director of EPA's Health Effects Research Laboratory in North Carolina.

Research Exposes Danger

Twenty years ago, many health standards were based on isolated, large doses of a single pollutant. A big dose of a chemical solvent, for example, could sicken or asphyxiate a worker. Two decades of research have taught scientists that chronic exposures even at extremely low levels can produce irreversible health problems. For example, prolonged exposure even to tiny quantities of ozone, a major component of smog, can cause serious lung damage. One study showed a few hours of exposure to a level of ozone smaller than the amount in the air on a bad day in Los Angeles can result in chest pain, coughing, and nausea.

Researchers have learned that some segments of the population—notably the young and the elderly—are especially susceptible to pollutants, and therefore that exposure standards must be calculated accordingly. Nitrates from fertilizers that have infiltrated groundwater pose a greater health risk to infants than to adults, for instance. And current air pollutant standards for sulfur dioxide are set to protect asthmatics, who are particularly sensitive to that chemical.

Over the past 20 years, scientists also have honed their ability to predict through laboratory experiments the effects of some chemicals on human health. In many cases, regulators no longer have to wait for sickness or death to strike before limiting exposure. The chemical 1,3-butadiene, widely used in synthetic rubber and plastic pipes, was considered safe until studies showed it produced cancer in laboratory animals. (It is now thought likely to be carcinogenic also in humans.)

Despite such progress, enormous gaps remain in the body of scientific knowledge. Research on the combined effect of exposures to several pollutants is almost nonexistent. What's more, the bulk of health-based environmental standards treat single pollutants in individual media: A standard for lead in air, for example, does not take into account that a person also is exposed to lead in drinking water and food. Nor do scientists understand the synergistic action of chemicals—the ways in which they combine or break down to form different, possibly far more potent, substances inside the body.

Is America a riskier place to live today than it was in 1970? Some vital signs say no. Within the past 20 years, for instance, U.S. life expectancy has increased by 4.4 years for men and 3.6 years for women. But if risk is measured by the quantity of toxic chemicals streaming into the environment, the hazards are greater than ever. "The levels of chemical usage and production have increased dramatically in the past 20 years," says Jerry Poje, staff toxicologist for the National Wildlife Federation, "and they are projected to increase even more in years to come."

More than a ton of toxic waste is produced each year for every man, woman, and child in the country, says the EPA. Of that amount, 135 billion pounds are discharged into the nation's waterways, while billions more are spewed into the air. In 1987, for instance, nearly 3 billion pounds of airborne pollutants were released. That figure accounts for only about half of the nation's top 30,000 industrial polluters.

At least 2,000 Americans die each year from cancer attributed to polluted air, according to government figures. Recently scientists have identified new sources of airborne pollution, such as radon, a naturally occurring radioactive gas. Radon levels in some areas are so high that the gas has been ranked as the second-leading cause of lung cancer after cigarette smoking. Fumes from household products and building materials also pose a serious threat, and in some buildings—including the EPA's headquarters in Washington, D.C.—workers have become ill from breathing the indoor air. Meanwhile, ozone pollution reached unhealthful levels in many cities in recent summers, and experts predict the problem will only grow worse.

Inadequate Protection

As Americans have witnessed one environmental calamity after another during the past two decades, legislators have responded with a flood of new laws and regulations. For example, widespread reports of illegal dumping of hazardous wastes triggered passage of the Toxic Substances Control Act in 1976. Similarly, the discovery of toxic chemicals seeping from the ground in the Love Canal neighborhood in Niagara Falls in 1977 prompted the 1980 Superfund law, which was created to clean up the worst toxic dump sites.

With legislation has come some success. The ban on DDT in 1972 has begun lowering levels of that pesticide in humans and wildlife populations. Laws requiring unleaded gasoline and catalytic converters in new cars starting in 1975 greatly improved air quality. And the decision in 1977 to prohibit manufacture and sale of PCBs has reduced that threat. But, as Ellen Silbergeld points out, these victories came about only on the strength of government-imposed bans. "Will it take that kind of intervention to make a difference with other chemicals?" she asks.

Critics argue that the government fails to protect the American public from many risks because it allows large numbers of new products to enter the environment without first examining the health consequences. David Doniger, an attorney with the Natural Resources Defense Council in Washington, D.C., contends the government should test new chemicals just as it tests new drugs. "It's perfectly okay [legally] to fill the environment and expose people to these chemicals without knowing

they're dangerous," he says. "It's as though ignorance is equated with safety."

Statutes often pay lip service to a more precautionary approach, and in some cases courts have forced the federal government to adopt controls. But the high cost of testing (borne by taxpayers, not producers), along with powerful resistance from the chemical industry, has effectively blocked implementation of a screening policy. Nowhere is that more apparent than in the area of pesticides.

Around 2.6 million pounds of pesticides are applied to fields each year in this country, to say nothing of the millions of pounds of other chemicals applied to crops, lawns, and gardens. Pesticides, such as insecticides and herbicides, are the only toxins purposefully introduced into the environment, yet research on their effects is extremely limited. In 1987, the National Academy of Sciences estimated that pesticide contamination in food could be responsible for up to 20,000 cancer cases annually. Ironically, last summer the academy found that farmers could be just as productive using fewer chemicals.

"Essentially what we've done is allowed an unregulated, uncontrolled, and, in my opinion, unethical experiment to be performed on the American public and our ecosystem," says Jerry Poje. In California's San Joaquin Valley, the world's richest, most intensively cultivated farm belt, the "experiment" may have backfired.

Seven percent of all pesticides used in the United States are applied in the vicinity of McFarland (population 6,200), an area resident Connie Rosales calls a "toxic fishbowl." The community's well water is tainted with DBCP, a now-banned pesticide known to cause cancer. For reasons still not understood, 16 children in Mc-Farland have developed the disease. Since 1975, 9 of them have died, most between 1982 and 1985—a rate eight times higher than normal for a town McFarland's size. Says Rosales, "Our kids have been the canaries in the coal mines."

The Legacy of Love Canal

The people of the San Joaquin Valley aren't the only ones paying for the nation's legacy of chemical use. Pat Brown and her family were among the 1,000 who left their homes in Niagara Falls a decade ago when chemicals such as benzene, dioxin, and trichlorethylene were discovered seeping from a long-abandoned canal that had been used as a dump for almost 22,000 tons of chemical waste. Love Canal, named for its builder William Love, has come to symbolize the costs of industrial negligence. But it also illustrates the chasm between what science can offer in the way of assurances on the health effects of pollution and what it cannot.

State and federal governments conducted studies in the Love Canal neighborhood, but so far none has proven any link between exposure to the chemicals and cancer or other ailments. Unsatisfied with the official findings, Pat Brown conducted her own, unofficial study. She was shocked at what she learned.

Working with the Ecumenical Task Force of Niagara Falls, which helps communities cope with environmental crises, Brown surveyed ten households. She found every home has at least one, and as many as three, cancer victims. Children from those homes are now giving birth to children with health problems and birth defects. "But," Brown adds scornfully, "we have to call this housewife data."

In September 1988, New York's top health official announced that about 250 Love Canal homes could be reinhabited. Six months later, scientists discovered toluene and PCBs in the soil of an area thought to be free of contamination. The resettlement plan was put on hold. "We are careful not to say 'safe'" when referring to the area, explained James Melius, director of New York's Division of Occupational Health and Environmental Epidemiology. "One thing we are learning about chemicals is that if we think something is safe today, we may learn a few years later that we were wrong."

From *National Wildlife*, February/March 1990. Reprinted by permission.

Poisoned Drinking Water: One Town's Solution

By Bill Lawren

The nightmare started in 1983, when the Hickey family of Long Pine Road in Whately, Massachusetts, noticed mysterious rashes erupting on their children's skin. Having ruled out allergies and skin diseases, the Hickeys checked their water supply. Sure enough, their well water was contaminated, laced with high levels of Temik and ethylene dibromide—pesticides that were used liberally on nearby tobacco and potato fields.

For Whately (population 1,500), the bad dream was just beginning. Like many other communities across the nation, its citizens suddenly found themselves facing an environmental calamity. When the state of Massachusetts checked private wells supplying as many as 175 homes in the low-lying agricultural area of East Whately, it found virtually all were contaminated. So for the next three years, 175 families lined up at an elementary school in West Whately to fill water containers from the school's outdoor tap. (The water supply at

Water board chairman Paul Fleuriel shows off the fruits of Whately's labors. (Seth Resnick)

the school, located on a hilltop overlooking the Connecticut River Valley, was unaffected by lowland contamination.)

"One poor woman with a station wagon came every two days," Frank Marchand, chairman of the Whately Water Resource Protection Committee, told the Associated Press. "She filled up so many milk jugs they stretched from the front seat all the way to the back window." The health of the entire community was at stake, but nothing short of a miracle or a mountain of cash could bring its residents safe water. In a dramatic show of unity, they managed both.

Townspeople knew there was a deep

and plentiful aquifer, untapped and uncontaminated, lying beneath 50 feet of clay. But if each of the 175 families dug its own well to tap that aquifer, in the words of town water superintendent William Smith, it would amount to "175 puncture wounds in that clay layer"—wounds that could allow pesticides to leak down and contaminate the deeper aquifer.

At an emotional town meeting in 1986, residents passed a bond issue to pay for digging a single well to supply all of East Whately's families. That, along with grants from two state agencies and a one-time hookup fee of $3,500 to $5,000 for each family, gave the town the $4 million it needed for the new system, which began providing uncontaminated water in 1987.

Today, the long lines at the elementary school have vanished, and the use of pesticides has dropped dramatically. Paul Fleuriel, chairman of the Whatley Water Commission, says the new water system is "as close to a permanent solution as we have in this modern day." It's also an object lesson in the potential rewards of concerted community action.

From *National Wildlife*, February/March 1990. Reprinted by permission.

WHAT YOU CAN DO

Automobile batteries should be recycled, repaired, and reused. Acid and metals in old batteries pose a hazard due to their toxic and corrosive nature. Contact a local battery shop, listed in the yellow pages of the telephone book under "Batteries—Storage—Retail."

The Healthy Home, by Linda Mason Hunter

The Toxic Avenger Strikes at Corporate Polluters

By Michael Satchell

It is midnight as the tiny Zodiac inflatable craft bounces north through the chop of America's most famously filthy harbor. With the downtown Boston skyline glowing about 2 miles behind and the eerie lights of the Monsanto Company plant dead ahead, Marco Kaltofen steers into a creek, cuts the 40-horsepower Nissan outboard, and begins paddling toward a set of discharge pipes. The lone toxic avenger and the chemical conglomerate are old adversaries, and tonight Kaltofen is returning to sample its wastes.

Kaltofen, 29, who operates a chemical-analysis laboratory for the Boston-based National Toxics Campaign, began his particular brand of environmental activism some five years ago. He slips into plants disguised as a utility worker or zips around the waterfront, staging hit-and-run forays while armed with a pH meter to measure acidity, a chemical-analysis kit, empty bottles, plastic bags, and a notebook. Boston's low-tech toxic detective samples and analyzes soil, air, and liquid discharges on or near the property of suspected polluters. His newest weapon is a radio-controlled model airplane equipped to fly over smokestacks and gather air samples.

With results in hand, he often calls a press conference to publicize his findings. If the pollution is severe, he takes his evidence to local authorities, and he is credited with spurring several successful prosecutions. As an environmental guerrilla, his activism has been more creative and effective than that of other so-called monkey wrenchers. Some radicals hammer spikes into old-growth trees to thwart the chain saw, pour sand into bulldozer tanks, switch survey stakes at oil-exploration sites, and commit other potentially destructive, illegal acts of "ecotage." Others favor legally questionable, high-profile, risky stunts like scaling bridges and tall buildings to hang banners or confronting whaling ships on the high seas.

Remedial Embarrassment

"My strategy is simple," Kaltofen says. "I try to embarrass the companies badly enough so they are forced to deal with the situation." Monsanto has been one of his favorite targets since he began analyzing its discharges and publicizing the results in 1984. Spokesman Michael Ferrante admits past violations of a pollution permit by the firm at the 120-year-old plant but stresses that Monsanto is working hard on remedial action. "We are moving expeditiously, but

Marco would like to see it done sooner," Ferrante says. "So would we, but we have to follow the law."

The Dutch-born Kaltofen began his environmental activism in 1982, plunging into the kind of direct-action showboating that garners easy headlines and television footage but has little permanent value. He chained himself to truckloads of sludge, delivered barrels of waste to corporate headquarters and annual stockholders' meetings, and plugged discharge pipes at polluting companies. He has been arrested more than a dozen times and convicted once of trespassing. "At some point I realized that delivering hazardous materials to a company office was in itself an act of violence," he says. "So I switched tactics."

Recently, federal criminal-enforcement efforts have bogged down in bureaucratic inertia and infighting, and Washington is encouraging local authorities to step up prosecution of major polluters. Massachusetts in May [1989] launched a 33-member state strike force of scientists, lawyers, and undercover agents to crack down, although there are still some like Kaltofen who believe that civilians are more effective than government at the investigative side of enforcement.

Successful Investigation

His most successful thrust to date has been against American Cyanamid in Linden, New Jersey, where Kaltofen spent two months surreptitiously gathering pollution samples at the chemical plant, interviewing employees, and sifting through public records. Based on his files, the state attorney general's office launched an investigation, and in 1987 the company pleaded guilty to 37 criminal counts of polluting the Rahway River. American Cyanamid was fined $900,000, at the time the largest penalty ever assessed in a New Jersey environmental case, and agreed to make major antipollution improvements. "Without the information from Marco, we would not have targeted this company," says Robert Candido, the former chief of environmental prosecutions, who handled the case. "I think he does an outstanding job in terms of public service."

Like many activists, Kaltofen is driven by a sense of deep frustration. When it comes to cleanup, things never happen fast enough. Equally infuriating is the fact that much toxic dumping violates no law, further evidence to him that the nation's environmental priorities are askew. His Boston Harbor bailiwick, where President George Bush turned environmental issues on Massachusetts Governor Michael Dukakis and helped sink his campaign last fall, and where flounder are still pulled from the water bearing golf ball-size tumors, is a perfect example. "There's a bill in the state legislature to close the harbor to fishing," Kaltofen says. "Why not close the harbor to toxic dumping?"

Enviro-Cops
on the Prowl for Polluters

By Stephen J. Hedges

It was the FBI that drew the attention early one morning last June [1989] when 70 federal agents swarmed into Rocky Flats, a federal nuclear-weapons facility near Denver, and declared the place under criminal investigation. It is *always* the FBI that draws

attention, it seems. But marching alongside the G-men were 17 agents from a special force of 48 criminal investigators whose specialty is bringing polluters to justice—the E-men of the Environmental Protection Agency (EPA). The unit was set up seven

Federal agents train state investigators in environmental busts. (Darryl Heikes/*U.S. News and World Report*)

120

years ago as an experiment and has since won the backing of many environmentalists. But it has only a few ardent supporters within the EPA.

The E-men may provide the kind of prosecutorial punch the Bush administration, busily trying to build its environmental credentials, could use. In its first six years, the force has won 385 indictments and 279 convictions, with about 100 cases still to be tried. The targets have run the gamut from a Winchester, New Hampshire, leather-tanning company that illegally dumped solvents and greases into the Ashuelot River, to W. R. Grace & Co., which was charged with making false statements about chemical solvents used at its Woburn, Massachusetts, plant and concealing the illegal dumping of chemicals on nearby grounds. The company pleaded guilty to one count and was fined $10,000. Lately, the targets have included the government itself. The raid on the Department of Energy's Rocky Flats, a joint EPA-FBI fraud investigation focusing on the illegal dumping and burning of hazardous wastes, was the E-men's most ambitious effort so far. They have opened similar investigations into the Department of Energy's (DOE) uranium-refining center at Fernald, Ohio, and at its West Valley nuclear-waste site near Buffalo, New York. And more probes of once-secretive DOE facilities are soon to come. "It's no-holds-barred now," says Marty Wright, chief of the EPA's National Enforcement Investigations Center. "There's almost an open-door policy for us."

Midnight Dumpers

In a nation that produces 265 million metric tons of hazardous wastes a year, wastes that increasingly are lacing rivers, creek beds, woods, and fields, there is no shortage of work for the E-men. Agents say tougher regulations have actually spawned large numbers of midnight dumpers, companies that cannot afford to dispose of wastes properly. In Illinois, for instance, the manager of an industrial plant was caught dumping a 55-gallon drum of polychlorinated biphenyls (PCBs), a suspected carcinogen, at a forest preserve. He pleaded guilty. Larger companies are just as culpable. Last year, Ocean Spray Cranberries, Inc., pleaded guilty to spewing undertreated wastewater into the Nemasket River at Middleboro, Massachusetts.

Congress actually made fouling the environment a crime when it passed the 1899 Rivers and Harbors Act, but few prosecutors applied the act.

Most cases begin with tips from troubled employees, which means that the agents nearly always are dealing with crimes that have already been committed. There is not enough manpower to put agents undercover, and there is nothing simple in proving crimes the agents do learn about. Chemicals do not leave fingerprints; illegal substances quickly wash away or are absorbed by soil. When agents do find the goods, they often must don oxygen masks and protective "moon suits" to collect samples of deadly substances.

The deployment of criminal investigators to environmental cases has bucked a long tradition of punishing polluters with hefty civil fines. Congress actually made fouling the environment a crime when it passed the 1899 Rivers and Harbors Act, but few prosecutors applied the act. Even a wave of tough environmental laws passed in the 1970s spurred only "haphazard" environmental enforcement, according to Judson Starr, former head of the Justice Department's environmental enforcement unit. That changed in 1982, when Starr and colleagues at Justice and the EPA persuaded their bosses to fund criminal investigators with the necessary expertise to enforce the tougher laws. The prosecutors wanted not EPA technicians but street-tough detectives who knew how to collect evidence, follow leads, and make arrests.

State Side

The upstart E-men, with their law-and-order approach, made few friends at the EPA. Not all the agency's leaders were convinced that they should be in the law-enforcement business. The squad grew slowly, from an initial 23 investigators to 48 by 1986. Fifteen more slots are expected under the 1990 federal budget, which still would mean just a handful of agents in each of the EPA's ten regions. To bridge the gap, the E-men have offered training to state law-enforcement and environmental officers as well as to Drug Enforcement Administration and FBI agents. Bush's EPA administrator, William Reilly, has promised that the agency "will get tougher on enforcement, which is the cornerstone of EPA's environmental program. We expect to see even more activity in the future."

It is not clear whether that means even greater emphasis on enviro-cops' criminal investigations. There are still plenty of officials around Washington who think that the best environmental-enforcement work is done through the EPA's civil inspectors or individual states. Those arguments raise the stakes of high-profile cases like the Rocky Flats investigation. The agents are betting that a new administration, one that has promised a crackdown on environmental criminals, will back up its tough talk with action that could turn attention to the E-men for a change.

The Pioneers

By Jackey Gold

At 3M's original tape manufacturing plant in St. Paul, Minnesota, built in 1939, recycling ideas percolate up often. About as often, in fact, as the hot glue that bubbles in huge vats, glue used in the company's adhesive tape products. Plant employees are now trying to figure out how to get the last bit of gooey, green glue out of the 55-gallon steel drums delivered daily from a 3M plant in Illinois. "There's competition in the plant to design a way to totally empty all our drums," says plant manager Gene Saienga, shouting over loud bursts of hissing steam that keeps the room humid and the tape static-free. A plastic purple bucket, half full of adhesive, is sitting in a nearby drum. "It's like the last bit of honey that gets caught in the jar," adds Saienga, pointing to the bucket. "Half a pail a day may not seem like much, but five days a week, it adds up. And that's less raw material we have to make and ship, and less pollution created in the course of manufacturing."

Welcome to corporate environmental consciousness, Minnesota-style. In a state with 10,000 lakes and some of the Northland's best canoeing, moose hunting, and eelpout fishing, it's no surprise to find a large manufacturer paying more than just lip service to the environment. But even for Minnesota, 3M is exceptional. Along with a St. Paul neighbor, $700-million-in-sales

specialty chemical maker H. B. Fuller, the company is often cited by activists as one of the most environmentally conscientious around. Both have invested millions in pollution control equipment. And both are religious about trying to recycle virtually everything—from defective duct tape and industrial solvent to 55-gallon steel drums—all without being required to by either federal or state regulations.

Up close, though, the similarities end. Granted, both 3M and Fuller preach a brand of environmental preventive medicine, making it a priority to stay ahead of state and federal environmental regulations. But at 3M the impetus clearly comes from the factory floor and the 2,400 ad hoc employee teams that have so far made waste reduction possible, while at Fuller the direction comes from the top. "We know which way the parade is going," says Fuller's CEO and president, Tony Anderson. "If there's an environmental problem somewhere in your company, the sooner and quicker you get at it, the better off you'll be."

At 3M, the company's environmental awareness dates back at least to the early 1970s, when a flurry of national environmental legislation such as the Clean Air Act first hit the books. In 1975, the company launched a formal, in-house pollution reduction program, the first of its type and a rather radical idea for American industry at the time. "Most governmental legislation involves costly add-on devices to clean up the problem, whatever it may be," says Russell Susag, the company's director of environmental regulatory affairs. "Our idea was to get rid of pollution or reduce it as much as possible, in the first place."

So far, 3M's program has saved it $482 million worldwide. It has eliminated some estimated 527,300 tons of air, water, sludge, and solid waste pollutants from the company's production processes, as well as 1.6 billion gallons of wastewater. In a separate energy conservation program, the company has saved more than $650 million.

Now, along with its internal program, 3M plans to "get ahead of the game," as Susag puts it, by spending $150 million over three years to install thermal oxidizers to control air pollution at some of its U.S. and overseas plants. More recently, it has started unveiling products that it considers environmentally helpful, including a foam landfill cover that takes up far less dump space than dirt. The cover is used to blanket garbage for short periods of time

WHAT YOU CAN DO

Take drain cleaners, toilet-bowl cleaners, oven cleaners, mothballs, and other highly toxic or corrosive products to a hazardous-waste collection site or licensed hazardous-waste disposal facility.

The Healthy Home, by Linda Mason Hunter

when dumping will not take place—such as a holiday weekend—before it disintegrates.

That 3M's corporate culture is a hotbed for employee inventiveness doesn't hurt, of course. But unlike many other environmental do-gooders, 3M and Fuller don't offer their employees financial incentives to pollute less. "We try to stay away from monetary awards because we think they're a demotivator," says Saienga. "This is about making a better tape today than you did yesterday."

Family Activism

At H. B. Fuller, the environmental spirit clearly comes from the CEO's office. Fuller has been run by the Anderson family since 1941, when Tony Anderson's father, former Minnesota governor Elmer L. Anderson, bought up a majority share of company stock. Today, the elder Anderson is remembered across Minnesota for helping found Voyageur National Park, a pristine canoeing and fishing area on the Canadian border. Both he and Tony are ardent naturalists.

When Tony Anderson wants to take a hike—as he often does—he doesn't need to do anything more than stop in at his company's Willow Lake research and development lab, which sits on the border of a 115-acre nature preserve outside of St. Paul. Built in cooperation with county officials, neighborhood residents, and the local Sierra Club, the three-floor, 110,000-square-foot lab is set up to maximize energy efficiency. It does so through the use of solar power, recycled heat exhaust and underground aquifer, and the Minnesota earth itself, which insulates the building on three sides.

Anderson is proud of H. B. Fuller's work to help revive the Willow Lake site. Back in 1979, he remembers, "we got a tongue-lashing at a meeting with local residents about chemical companies and the waste they produce. I put together a committee of the Sierra Club, the state, the city, and neighbors. Two years of work and a quarter of a million dollars were saved because we didn't need an environmental impact statement."

> *It's part of the philosophy of the company not to use asbestos. It's a carcinogen. If it's bad news here, it's bad news everywhere.*

It was Tony Anderson, too, who pushed the company's board of directors to accept a worldwide environment, health, and safety policy in 1986. Conceived in the wake of the 1984 gas leak at Union Carbide's Bhopal, India, factory, which killed 3,598, the policy attempts to bring 100 H. B. Fuller plants, warehouses, and labs around the world up to the same strict environmental standards. Anderson became somewhat obsessed with the Bhopal accident, clipping everything he could find about it from newspapers and magazines. What disturbed him the most were news stories that said an almost identical Carbide plant in West Virginia had operated in a completely different way despite the same set of corporate safety standards.

Today, every Fuller facility, regardless of where it is housed, is required to list the hazardous materials it uses, as well as its handling methods. Anderson may have his work cut out for him: Three years ago, it became apparent that some of the company's foreign plants had no relevant local standards to follow. "In many countries, they don't even have environmental regulations, and if they do, they don't enforce them," Anderson says. And then there's Fuller's policy for emergency evacuations. It calls for such plans to be in use at each site by the end of this year [1990]—a goal Anderson thinks he will be able to meet.

Fuller's global brand of environmentalism hasn't come cheaply. A recent audit, for example, showed that the company's plant in Hamamatsu City, Japan, used asbestos to make an industrial adhesive. Back home, the Willow Lake lab staff designed an alternative for the product. But the company's customer in Japan refused to buy it because it preferred the original asbestos-filled adhesive and knew it could be bought elsewhere in Japan. Fuller bit the bullet and cut the product anyway, which resulted in an $800,000 annual sales loss. Says Joseph Pellish, director of regulatory services: "It's part of the philosophy of the company not to use asbestos. It's a carcinogen. If it's bad news here, it's bad news everywhere."

Back in the United States, Fuller's move to recycle raw materials and defective products, like 3M's, is going on with a vengeance. At the company's adhesives and sealant plant in Fridley, Minnesota, for example, any sealant removed in the cleaning of six high-sheer blade mixers (which work like giant taffy pullers) is recycled with new raw materials. "It's just like making a cake," chuckles plant manager Don Wiskow.

Well, not quite. But here is a company that often takes a distinctly down-home approach to recycling. Take Fuller's efforts to siphon off the talc, clay, and calcium carbonate dust it generates in the Fridley plant, which would otherwise be released into the air. In a process that has been used by the flour industry for years, Fuller takes the dust, some 200 pounds every three days, and recycles it. The real incentive, of course, is cost reduction. Still, "I came from Buffalo, where Love Canal was right up the street," says Wiskow. "Concern for ecology is definitely a part of why we do this."

Which may explain why there are no underground storage tanks in use at Fridley—or at most other Fuller plants, for that matter. Instead, the plant has three huge silos for holding the hydrocarbon oils it uses in production, and those silos are housed in a separate cinderblock vault large enough to contain a spill if one should occur.

A Continuing Effort

But Fuller isn't resting on its laurels. It plans, for example, to start separating its plastic waste from its paper waste later this year. "Half of what we're now throwing away can be recycled," says Wiskow.

Fuller is not just giving the public what it wants to hear. Back in 1983, when Wiskow first came to Fuller, he discovered nearly 1,000 barrels on company property. Before the barrels could be landfilled, Wiskow went through them all to ensure they didn't contain anything hazardous, a task that took a year and a half.

These days, every barrel of waste that leaves the Fridley plant is assigned a number and is documented for landfill purposes—assuming, of course, that it hasn't already been worked back into the production process.

The same kind of vigilance operates at Fuller's Monarch division plant in nearby Columbia Heights, which produces 10,000 gallons of chlorine-based cleaners a day (with names like "Iodine Teat Dip" and "Super Udder Wash") for the food-processing and dairy industries. Trouble was, a similar Fuller plant in Tulare, California, sprang a chlorine gas leak in 1987. While no one was hurt, the plant and some local residents had to be evacuated. This put the fear of chlorine gas into Mike Jensen, manager of the Columbia Heights plant. When Jensen heard that another, recently acquired Fuller plant was using a form of liquid chlorine instead of the chlorine gas, he convinced management to let him replace one with the other, even though the liquid was slightly more expensive. Today, all the Fuller plants use the same substitute.

Like 3M, H. B. Fuller has won raves for its position on the environment. In fact, the company has become a darling of social fund portfolio managers, despite the fact that fourth-quarter earnings came in below estimates and even though it has gotten negative publicity for one of its glues, which has become a favorite for drug-abusing teenagers in Latin America.

3M has been less fortunate. Last year the company was hit with the highest industry fine ever assessed by the Minnesota Pollution Control Agency for violating particulate standards at its hazardous waste incinerator in suburban Cottage Grove, just south of St. Paul. This came on top of a $95,000 fine the company paid in 1988 for exceeding "opacity" limits (the density of smokestack emissions) at the same site. Then there was a 1987 legal settlement in which 3M paid local environmental groups and the U.S. government a total of $158,000 for spewing more-than-allowable amounts of oil, grease, and bacteria out of the Cottage Grove wastewater treatment plant.

When asked about these problems, 3M CEO Allen F. Jacobson responds, "It's true we've had some difficulties, but we've corrected them. We intend to operate our facilities in the most environmentally responsible manner possible." In the bigger scheme of things, of course, there isn't much doubt that both 3M and H. B. Fuller are imbued with a strong sense of concern for their surroundings. For 3M's Jacobson, it's "really as much about community relations" as it is about protecting the company from possible liability. And for Fuller's Anderson, it's about being a Minnesota company. "We've grown up with a respect for the environment," he emphasizes. It is a respect the rest of the American population is quickly learning as well.

From *Financial World*, January 23, 1990. Reprinted by permission.

TROPICAL FORESTS

Tropical forests teem with life of all kinds, including human. These two young men are members of the Penan tribe of Borneo. Like other inhabitants of rainforests from Africa to Brazil, the Penan are fighting a difficult battle to save their land and their way of life. (John Werner, Endangered People Project)

PLAYING WITH FIRE

By Eugene Linden

The skies over western Brazil will soon be dark both day and night. Dark from the smoke of thousands of fires, as farmers and cattle ranchers engage in their annual rite of destruction: clearing land for crops and livestock by burning the rainforests of the Amazon. Unusually heavy rains have slowed down the burning this year, but the dry season could come at any time, and then the fires will reach a peak. Last year the smoke grew so thick that Porto Velho, the capital of the state of Rondonia, was forced to close its airport for days at a time. An estimated 12,350 square miles of Brazilian rainforest—an area larger than Belgium—was reduced to ashes. Anticipating another conflagration this year, scientists, environmentalists, and TV crews have journeyed to Porto Velho to marvel and despair at the immolation of these ancient forests.

After years of inattention, the whole world has awakened at last to how much is at stake in the Amazon. It has become the front line in the battle to rescue earth's endangered environment from humanity's destructive ways. "Save the rainforest," long a rallying cry for conservationists, is now being heard from politicians, pundits, and rock stars. The movement has sparked a confrontation between rich industrial nations, which are fresh converts to the environmental cause, and the poorer nations of the Third World, which view outside interference as an assault on their sovereignty.

Some of the harshest criticism is aimed at Brazil. The largest South American country embraces about half the Amazon basin and, in the eyes of critics, has shown a reckless penchant for squandering resources that matter to all mankind. Government leaders around the world are calling on Brazil to stop the burning. Two delegations from the U.S. Congress, which included Sen. Albert Gore, Jr., of Tennessee and Sen. John Chafee of Rhode Island, traveled to the Amazon earlier this year [1989] to see the plight of the rainforest firsthand. Says Gore: "The devastation is just unbelievable. It's one of the greatest tragedies of all history."

The vast region of unbroken green that

*I*t's dangerous to say the forest will disappear by a particular year, but unless things change, the forest *will* disappear.

surrounds the Amazon River and its tributaries has been under assault by settlers and developers for 400 years. Time and again, the forest has defied predictions that it was doomed. But now the danger is more real and imminent than ever before, as loggers level trees, dams flood vast tracts of land, and gold miners poison rivers with mercury. In Peru the forests are being cleared to grow coca for cocaine production. "It's dangerous to say the forest will disappear by a particular year," says Philip Fearnside of Brazil's National Institute for Research in the Amazon, "but unless things change, the forest *will* disappear."

That would be more than a South American disaster. It would be an incalculable catastrophe for the entire planet. Moist tropical forests are distinguished by their canopies of interlocking leaves and branches that shelter creatures below from sun and wind, and by their incredible variety of animal and plant life. If the forests vanish, so will more than 1 million species—a significant part of earth's biological diversity and genetic heritage. Moreover, the burning of the Amazon could have dramatic effects on global weather patterns—for example, heightening the warming trend that may result from the greenhouse effect. "The Amazon is a library for life sciences, the world's greatest pharmaceutical laboratory, and a flywheel of climate," says Thomas Lovejoy of the Smithsonian Institution. "It's a matter of global destiny."

To Brazilians, such pressure amounts to unjustified foreign meddling and a blatant effort by the industrial nations to preserve their economic supremacy at the expense of the developing world. Brazilian President Jose Sarney has denounced the criticism of his country as "unjust, defamatory, cruel, and indecent." How can Brazil be expected to control its economic development, he asks, when it is staggering under a $111 billion foreign debt load? By what right does the United States, which spews out more pollutants than any other nation, lecture poor countries like Brazil on their responsibilities to mankind?

Yet Sarney is caught between conflicting, and sometimes violent, forces within his nation. On one side are the settlers and developers, often backed by corrupt politicians, who are razing the forests to lay claim to the land. On the other are hundreds of fledgling conservation groups, along with the Indian tribes and rubber tappers whose way of life will be destroyed if the forests disappear. The clash has already produced the world's most celebrated environmental martyr, Chico Mendes, a leader of the rubber tappers who was murdered for trying to stand in the way of ranchers.

The passions behind the fight are easy to understand for anyone who has seen the almost unimaginable sweep of the Amazon basin. The river and forest system covers 2.7 million square miles (almost 90 percent of the area of the contiguous United States) and stretches into eight countries beside Brazil, including Venezuela to the north, Peru to the west, and Bolivia to the south. An adventurous monkey could climb into the jungle canopy in the foothills of the Andes and swing through 2,000 miles of continuous 200-foot-high forest before reaching the Atlantic coast. The river itself, fed by more than 1,000 tributaries, meanders for 4,000 miles, a length second only to the Nile's 4,100 miles. No other river compares in volume: Every hour the Amazon delivers an average of 170 billion gallons of water to the Atlantic—60 times the flow of the Nile. Even 1,000 miles upriver, it is often impossible to see from one side of the Amazon to the other.

The Living Jungle

The jungle is so dense and teeming that all the biologists on earth could not fully describe its life forms. A 1982 U.S. National Academy of Sciences report estimated that a typical 4-square-mile patch of rainforest may contain 750 species of trees, 125 kinds of mammals, 400 types of birds, 100 kinds of reptiles, and 60 amphibians. Each type of tree may support more than 400 insect species. In many cases the plants and animals assume Amazonian proportions: lily pads that are 3 feet or more across, butterflies with 8-inch wing spans, and a fish called the pirarucu, which can grow to more than 7 feet long. Amid the vast assortment of jungle life, creatures command every trick in nature's book to fool or repel predators, attract mates, and grab food. Caterpillars masquerade as snakes, plants exude the smell of rotting meat to attract flies as pollinators, and trees rely on fish to distribute their seeds when the rivers flood.

But the diversity of the Amazon is more than just good material for TV specials. The rainforest is a virtually untapped storehouse of evolutionary achievement that will prove increasingly valuable to mankind as it yields its secrets. Agronomists see the forest as a cornucopia of undiscovered food sources, and chemists scour the flora and fauna for compounds with seemingly magical properties. For instance, the piquia tree produces a compound that appears to be toxic to leaf-cutter ants, which cause millions of dollars of damage each year to South American agriculture. Such chemicals promise attractive alternatives to dangerous synthetic pesticides. Other jungle chemicals have already led to new treatments for hypertension and some forms of cancer. The lessons encoded in the genes of the Amazon's plants and animals may ultimately hold the key to solving a wide range of human problems.

Scientists are concerned that the destruction of the Amazon could lead to climatic chaos. Because of the huge volume of clouds it generates, the Amazon system plays a major role in the way the sun's heat is distributed around the globe. Any disturbance of this process could produce far-reaching, unpredictable effects. Moreover, the Amazon region stores at least 75 billion tons of carbon in its trees, which when burned spew carbon dioxide (CO_2) into the atmosphere. Since the air is already dangerously overburdened by carbon dioxide from the cars and factories of industrial nations, the torching of the Amazon could magnify the greenhouse effect—the trapping of heat by atmospheric CO_2. No one knows just what impact the buildup of CO_2 will have, but some scientists fear that the globe will begin to warm up, bringing on wrenching climatic changes.

A Popular Cause

As the potential consequences of rainforest destruction became more widely known, saving the Amazon became the cause of 1989. In New York City, Madonna helped organize a benefit concert called Don't Bungle the Jungle, which also featured the B–52s and the Grateful Dead's Bob Weir. Xapuri, the remote town where Mendes lived and died, has been besieged by journalists, agents, and pilgrims. Robert Redford, David Puttnam, and other prominent moviemakers have sought the rights to film the Mendes story.

In the face of pressure from abroad and complaints from environmentalists at home, Brazil has grudgingly begun to respond. In April [1989], only a few months after denouncing the environmental movement as a foreign plot to seize the forests, the Sarney administration announced a hastily patched-together conservation package dubbed Our Nature. Much of the language was ambiguous, but the program contained promising provisions, such as the temporary suspension of tax incentives that spur the most wasteful forest exploitation. Says Celio Valle, director of ecosystems at the government's newly created environmental agency: "Before, we used to consider Brazilian environmental groups as the enemy, but now we consider them allies." Amazonian development may become a significant issue in this fall's [1989] presidential campaign. Fernando Collor de Mello, a member of the conservative National Reconstruction Party and a leading candidate to succeed Sarney, has said he believes in preserving the forests, though critics doubt his sincerity.

Many Brazilians still believe the Amazon is indestructible—a green monster so huge and vital that it could not possibly disappear. Asked about a controversial hydroelectric project that might flood an area as large as Britain, a Brazilian engineering consultant said, "Yes, that's a big area, but in terms of the Amazon it's small." Maintained Sarney recently: "It's not easy to destroy a rainforest. There are recuperative powers at work."

Yet the rainforest is deceptively fragile. Left to itself, it is an almost self-sustaining ecosystem that thrives indefinitely. But it does not adapt well to human invasions and resists being turned into farmland or ranchland. Most settlers find that the lush promise of the Amazon is an illusion that vanishes when grasped.

The forest functions like a delicately balanced organism that recycles most of its nutrients and much of its moisture. Wisps of steam float from the top of the endless palette of green as water evaporates off the upper leaves, cooling the trees as they collect the intense sunlight. Air currents over the forest gather this evaporation into clouds, which return the moisture to the system in torrential rains. Dead animals and vegetation decompose quickly, and the resulting nutrients move rapidly from the soil back to growing plants. The forest is such an efficient recycler that virtually no decaying matter seeps into the region's rivers.

But when stripped of its trees, the land becomes inhospitable. Most of the Amazon's soil is nutrient-poor and ill-suited to agriculture. The rainforest has an uncanny capacity to flourish in soils that elsewhere would not even support weeds.

Dreamers and Schemers

Throughout history, would-be pioneers and developers have discovered just how unreceptive the Amazon can be. Henry Ford tried twice to carve rubber empires out of the rainforest in the 1920s and 1930s. But when the protective canopy was cut down, the rubber trees withered under the assault of sun, rain, and pests. In 1967 Daniel Ludwig, an American billionaire, launched a rashly ambitious project to clear 2.5 million acres of forest and plant gmelina trees for their timber. He figured that the imported species would not be susceptible to Brazil's pests. Ludwig was wrong, and as his trees died off, he bailed out of the project in 1982.

The Brazilian government, meanwhile, came up with development schemes of its own. In the early 1970s the country built the Trans-Amazon Highway, a system of roads that run west from the coastal city of Recife toward the Peruvian border. The idea was to prompt a land rush similar to the pioneering of the American West. To encourage settlers to brave the jungle, the government offered transportation and other incentives, allowing them to claim land that they had "improved" by cutting down the trees.

But for most of the roughly 8,000 families that heeded the government's call between 1970 and 1974, the dream turned into a bitter disappointment. The soil, unlike the rich sod in the western United States, was so poor that crop yields began to deteriorate badly after three or four years. Most settlers eventually gave up and left.

Yet the failed dreams of yesterday have not discouraged Brazil from conjuring up more grand visions for today. The country has continued to build roads, dams, and settlements, often with funding and technical advice from the World Bank, the European community, and Japan. Two of the largest—and, to the rainforest, most threat-ening—projects are Grande Carajas, a giant development program that includes a major mining complex, and Polonoroeste, a highway-and-settlement scheme.

The $3.5 billion, 324,000-square-mile Grande Carajas Program, located in the eastern Amazon, seeks to exploit Brazil's mineral deposits, perhaps the world's largest, which include iron ore, manganese, bauxite, copper, and nickel. The principal iron-ore mine began production in 1985, and its operation has little impact on the forest. The problem, however, is the smelters that convert the ore into pig iron. They are powered by charcoal, and the cheapest way to obtain it is by chopping down the surrounding forests and burning the trees. Environmentalists fear that Grande Carajas will repeat the dismal experience of the state of Minas Gerais in southeastern Brazil, where pig-iron production consumed nearly two-thirds of the state's forests.

In the other huge project, Polonoroeste, the government is trying to develop

---

WHAT YOU CAN DO

The Nature Conservancy has formed an "Adopt-an-Acre Program" that enables individuals to buy and protect specific areas of threatened rainforest. For $30, you can adopt one acre of tropical land. In return, you'll receive an honorary land deed, information about the protected area, and semiannual letters from a local conservationist involved in managing the area. For more info, contact the Nature Conservancy, 1815 N. Lynn St., Arlington, VA 22209 or call (800) 872-1899.

the sprawling western state of Rondonia. The program, backed by subsidies and built around a highway through the state called BR–364, was designed to relieve population pressures in southern Brazil. But Polonoroeste has made Rondonia the area where rainforest destruction is most rapid, and the focal point of the fight to save the Amazon.

The results of the development have been chaotic and in some cases tragic. Machadinho, for instance, was supposed to be a model settlement village with gravel roads, schools, and health clinics. But when a surge of migrants traveled down BR–364 to Machadinho in 1985, orderly development became a pell-mell land grab. Settlers encountered the familiar scourges of the rainforest: poor soil and inescapable mosquito-borne disease. Decio Fujizaki, a settler who came west four years ago, has just contracted malaria for the umpteenth time. Says he: "I always wanted my own plot of land. If only it wasn't for this wretched disease."

Instead of model settlements, the Polonoroeste project has produced impoverished itinerants. Settlers grow rice, corn, coffee, and manioc for a few years, until the meager soil is exhausted, then move deeper into the forest to clear new land. The farming and burning thus become a perpetual cycle of depredation. Thousands of pioneers give up on farming altogether and migrate to the Amazon's new cities to find work. For many, the net effect of the attempt to colonize Rondonia has been a shift from urban slums to Amazonian slums. Says Donald Sawyer, a demographer from the University of Minas Gerais: "The word is out that living on a 125-acre plot in the jungle is not that good."

The abandoned fields wind up in the hands of ranchers and speculators who have access to capital. Thanks to tax breaks and subsidies, these groups can often profit from the land even when their operations lose money. According to Roberto Alusio Paranhos do Rio Branco, president of the Business Association of the Amazon, nobody would farm Rondonia without government incentives and price supports for cocoa and other crops.

Rondonia's native Indians have fared worse than the settlers. Swept over by the land rush, one tribe, the Nambiquara, lost half its population to violent clashes with the immigrants and newly introduced diseases like measles. Jason Clay, director of research for Cultural Survival, an advocacy organization for the Indians, says that when the Nambiquara were relocated as part of Polonoroeste, the move severed an intimate connection, forged over generations, to the foods and medicines of their traditional lands. That deprived them of their livelihood and posterity of a wealth of information about the riches of the forest. Says Clay: "Move a hunter-gatherer tribe 50 miles, and they'll starve to death."

Forest Loss Continues

Amid the suffering of natives and settlers, the one constant is that deforestation continues. Since 1980 the percentage of Rondonia covered by virgin forest has dropped from 97 percent to 80 percent. Says Jim LaFleur, an agricultural consultant with 13 years' experience working on colonization projects in Rondonia: "When I fly over the state, it's shocking. It's like watching a sheet of paper burn from the inside out."

A similar debacle could occur in the western state of Acre. It is still virtually pristine, having lost only 4 percent of its forests, but the rate of deforestation is increasing sharply as cattle ranchers expand their domain. Development in Acre has sparked a series of bloody confrontations between ranchers and rubber tappers, who want to preserve the forests so they can save their traditional livelihood of harvesting latex and Brazil nuts. It was this conflict that killed Mendes.

This courageous leader did not set out to save the Amazon but to improve the lot of rubber tappers, or *seringueiros*. He and his men would try to dissuade peasants from clearing land. The ranchers were eager to get rid of him, but he survived one assassination attempt after another. The conflict finally came to a head last year, when Mendes confronted a rancher named Darli Alves da Silva, who wanted to cross land claimed by rubber tappers to cut an adjacent 300-acre plot. After Mendes and a group of 200 seringueiros peacefully turned back the rancher and 40 peons, death threats against him grew more frequent. In December he was killed with a shotgun as he stepped out of his doorway. Alves and two of his sons were convicted of the murder but have appealed the verdict.

Mendes became a hero to environmentalists not only because he fought and died to stop deforestation but also because of the way of life he was defending. The rubber tappers are living proof that poor Brazilians can profit from the forest without destroying it. According to Stephan Schwartzman of the Environmental Defense Fund, seringueiros achieve a higher standard of living by harvesting the forest's bounty than do farmers who cut the forest and plant crops.

One of Mendes' most important achievements was to help convince the Inter-American Development Bank to suspend funding temporarily for further paving of BR–364 between Rondonia and Acre. But the Brazilian government is again seeking the $350 million needed to complete the road all the way to Peru, a prospect that alarms environmentalists. "One lesson we have learned in the Amazon is that when you improve a road, you unleash uncontrolled development on the rainforest," says John Browder, a specialist on Rondonia's deforestation from Virginia Polytechnic Institute.

Among other things, environmentalists fear that completion of the road will provide entree for Japanese trading companies that covet the Amazon's vast timber resources. Acre's governor, however, argues that the road is needed to end the state's isolation and claims that the state will not repeat the mistakes of Rondonia.

The debate over the Acre road places environmentalists in an uncomfortable position, essentially telling Brazilians that they cannot be trusted with their own development. Raimundo Marques da Silva, a retired public servant who helped build Acre's original dirt highway, asks, "How would Americans feel if years ago we had told them they could not build a road from New York to California because it would destroy their forests?"

Still, some Brazilians do accept that the outside world has a legitimate interest in the Amazon. Jose Lutzenberger, an outspoken environmentalist, notes that the Brazilians trying to develop the rainforest are themselves outsiders to the area. "This talk of 'We can do with our land what we want' is not true," he says. "If you set your house on fire it will threaten the homes of your neighbors."

If the rainforest disappears, the process will begin at its edges, in places such as Acre and Rondonia. While the Amazon forest as a whole generates roughly half of its own moisture, the percentage is much higher in these western states, far from the Atlantic. This means that deforestation is likely to have a more dramatic impact on the climate in the west than it would in the east. "Imagine the effects of a dry season extended by two months," says Fearnside, of Brazil's National Institute for Research in the Amazon. The process of deforestation could become self-perpetuating as heat, drying, and wind cause the trees to die on their own.

Exploring Alternatives

This does not have to happen. A dramatic drop in Brazil's birth rate promises to reduce future pressures to cut the forests, and experts believe the country could halt much of the deforestation with a few actions. By removing the remaining subsidies and incentives for clearing land, Brazil could both save money and slow the speculation that destroys the forests. Many environmentalists prefer this approach to the enactment of new laws. Brazilians have developed a genius, which they call *jeito*, for getting around laws, and many sound environmental statutes on the books are ignored.

The government could also stop some of the more wasteful projects it is currently

planning. Part of the problem in the Amazon has been ill-conceived plans for development that destroy forests and drive the country deeper into debt. Most hydroelectric dams, for example, have proved unsuitable in the region. The Balbina Dam, which was completed in 1987 and began operating early this year [1989], flooded a huge area at great cost to produce relatively little power. It killed trees, poisoned fish, and provided breeding grounds for billions of malarial mosquitoes. Despite this experience, the government plans to build scores of additional dams.

Fabio Feldmann, the leading environmentalist in the Brazilian congress, alleges that much of the momentum behind the dam projects and other large public works derives from an extremely lucrative relationship between the major contractors and politicians. A dam may not have to make all that much sense if it generates sufficient *commissao* (commissions) for the right people.

---

*I*f the burning of the forests goes on much longer, the damage may become irreversible.

Perhaps the best hope for the forests' survival is the growing recognition that they are more valuable when left standing than when cut. Charles Peters of the Institute of Economic Botany at the New York Botanical Garden recently published the results of a three-year study that calculated the market value of rubber and exotic products like the aguaje palm fruit that can be harvested from the Amazonian jungle. The study, which appeared in the British journal *Nature,* asserts that over time, selling these projects could yield more than twice the income of either cattle ranching or lumbering.

But if the burning of the forests goes on much longer, the damage may become irreversible. Long before the great rainforests are destroyed altogether, the impact of deforestation on climate could dramatically change the character of the area, lead to mass extinctions of plant and animal species, and leave Brazil's poor to endure even greater misery than they do now. The people of the rest of the world, no less than the Brazilians, need the Amazon as a functioning system, and in the end, this is more important than the issue of who owns the forest. The Amazon may run through South America, but the responsibility for saving the rainforests, as well as the reward for succeeding, belongs to everyone.

From Time, September 18, 1989. Copyright © 1989 The Time, Inc., Magazine Co. Reprinted by permission.

 EARTH CARE ACTION

What Can Americans Do?

By Michael D. Lemonick

It is easy for Americans to criticize Brazil's record on the environment, since they already live in a rich, industrialized country. But the United States achieved this status largely by doing just what Brazil is condemned for: ruthlessly exploiting natural resources—including cutting down most of its native forests—and fouling the environment in the process. Why, ask Brazilians, should they forgo the benefits of development just because North Americans have suddenly got religion?

Even more galling, the United States continues to be a major degrader of the planet. Its cars and factories pump hundreds of millions of tons of chemicals into the air each year, contributing to such atmospheric evils as greenhouse warming and acid rain. No wonder Brazil cries "environmental imperialism."

If Americans are truly interested in saving the rainforests, they should move beyond rhetoric and suggest policies that are practical—and acceptable—to the understandably wary Brazilians. Such policies cannot be presented as take-them-or-leave-them propositions. If the United States expects better performance from Brazil, Brazil has a right to make demands in return. In fact, the United States and

Brazil need to engage in face-to-face nego-tiations as part of a formal dialogue on the environment between the industrial na-tions and the developing countries. The two sides frequently negotiate on debt re-financing and other issues. Why not put the environment on the agenda?

To get developing countries like Brazil to talk seriously, the United States might have to take some unilateral steps. It is not so much that the Brazilians care particu-larly whether Los Angeles is smoggy or Akron acrid. But a willingness by Ameri-cans to make painful choices in atoning for their own sins would go far to defuse the Brazilians' indignation. Further stiffening of fuel-economy standards for new Ameri-can cars, for example, would send a strong signal. So would an increase in federal gasoline taxes to bring U.S. fuel prices closer to those in Brazil and the rest of the world. And perhaps most to the point, the United States should stop its questionable logging of ancient forests in the Pacific Northwest and Alaska.

At the bargaining table, the United States, Western Europe, and Japan would have a huge carrot to offer: debt relief. De-veloping countries might be more willing to curb environmental abuses if part of their $1.3 trillion foreign debt were for-given. There is precedent for this strategy in the so-called debt-for-nature swaps pio-neered in Bolivia and Costa Rica. In these plans, nations have received debt reduc-tions if they have agreed to protect certain tracts of land from development.

But Brazilians have resisted such swaps, fearing a loss of sovereignty over lo-cal resources. Instead of offering debt relief in return for nature preserves that some Brazilians do not want, the United States could offer it in exchange for something the Brazilians need: responsible development. The forest land should be utilized without being destroyed. A dam that flooded a vast area to produce small amounts of electric-ity would not qualify for debt relief, for ex-ample; a well-managed tree-harvesting op-eration would.

Another area in which Brazil could use help is in the training of local conservation-ists, who lag far behind their American counterparts in expertise, equipment, and financing. Says Fernando Cesar Mesquita, head of Ibama, the Brazilian environmental protection agency: "We have created 130 conservation areas, and we have desig-nated 2 million hectares of national park-land, but we need money to buy that land, put up fences, and administer these parks."

Other, better ideas might come from direct negotiations. The talks would be the crucial step in preserving the Amazon re-gion. By agreeing to discuss the situation and offering reasonable suggestions, rather than simply preaching, the United States would at least have a chance of doing what it has so far failed to do: Nudge Brazil to-ward a more environmentally sound devel-opment policy.

Iguana Mama

By Noel Vietmeyer

Last September [1988], when scientist Dagmar Werner first ate one of the green iguanas she had been raising, she found the fruit of her labor "something like a goulash," and pronounced it "very delicious."

For the German-born herpetologist, there was another taste of victory, however, and it was even better than that of the light white meat cooked in a coconut sauce. The meal marked the end of the first cycle of an improbable animal husbandry experiment in which Werner hatched iguana eggs in captivity, kept the iguanas caged through their vulnerable first months, and then released them to farmers' backyards in Panama.

Werner's work promises to help save not only the threatened lizard—considered a great delicacy throughout its range—but at least some of Central America's vanishing tropical forest. If farmers can be convinced to go into the iguana business, they will have every reason to preserve their livestock's treetop habitat.

So far, Werner's success with the experiment, which started in 1983, has wildly exceeded the prediction of experts. "In five years, she domesticated an animal we thought would take a century to learn to manage," says Ira Rubinoff, director of the Smithsonian Tropical Research Institute in Panama.

The 45-year-old scientist has done it with panache that goes far beyond an aptitude for research. In her standard jeans and T-shirt, Werner looks more like a farmhand than a scientist, a symbol perhaps of what it takes to succeed with such a bizarre scheme. "Dagmar is aggressive, a go-getter, and a little nuts," says National Zoo curator Dale Marcellini. "Just the person for the task."

Six years ago, when the Smithsonian Tropical Research Institute signed on Werner to direct its new Iguana Management Project, the task seemed especially daunting. According to 35 experts who attended an institute workshop that spring, the trial-and-error required for ranching a new species would take a score of scientists decades to yield useful results.

Despite the bleak outlook, it seemed an effort well worth making. Central American farmers can coax crops out of the thin, acidic soil for only about three years. Then they turn to cattle raising for another decade at most, until erosion generally makes the land useless. An acre of the poor grazing land yields about a dozen pounds of meat annually, estimates Werner. "That same land," she says, "if left as forest and given over to raising iguanas, could produce more than 200 pounds of iguana meat a year."

One of the most widespread creatures in South America, the green iguana occurs

Known as "chicken of the trees," green iguanas, which can weigh 10 pounds each, have long been an important source of protein in Central America, but their populations have plummeted in the wild. (Karen M. Asis)

*A*lthough it looks something like a prehistoric dragon with its claws and spines, the iguana is a shy herbivore whose main pleasures in life are nibbling on tender treetops and basking in the sun.

in lowland forests from Mexico to Brazil. Although it looks something like a prehistoric dragon with its claws and spines, it is a shy herbivore whose main pleasures in life are nibbling on tender treetops and basking in the sun, where the heat aids its cold-blooded digestion. For a lizard, it is a giant, and can grow to 6 feet or more—though about half its length is a tail only the thickness of a whip.

A big iguana can weigh more than 10 pounds. That is a nice size for a family dinner, and throughout Latin America people love to eat iguanas. The rich consider them a tasty treat. In markets, customers willingly pay more for iguana meat than for fish, poultry, pork, or beef. When the lizards were more plentiful, they were a food staple for the malnourished poor, who hunted them with slingshot, trap, and noose. Some even used trained "iguana dogs" that run down the creatures and hold them until hunters catch up. At Werner's first harvest, villagers used another time-honored method. One climbed a tree and shook branches while others captured the falling prey.

Lizards have been important to people since prehistoric times. In the mid–1500s they were "a most remarkable and wholesome food," according to one Spanish conquistador in Yucatan who also wrote, "There are so many of them that they are a great assistance to everyone during Lent." Iguana eggs are also tasty, and according to folklore, enhance sexual potency. Iguana fat is prized as a salve for burns, cuts, and sore throats.

Twenty years ago, iguanas—known throughout Central America as "chicken of the trees"—were so plentiful they were commonly taken to market by the truckload. There, vendors sat beside heavy iron pots suspended over flames, selling spicy iguana stew. Trussed-up live lizards waited to be sold nearby; customers hauled the creatures away in full gunnysacks and wicker baskets. Now, on the black market, a single animal can cost $20 (U.S.) or more, up from 80 cents in 1976 on the open market and $4.80 in 1979. The price, says Werner, "is like a thermometer for measuring how endangered the animal is."

Scientists don't know exactly when the decline started or how severe it has become, but there is no question that the green iguana faces extinction in many parts of Central America. "Everywhere I go, everyone says the iguana is gone," says Werner. Many countries classify it as threatened; some, like Panama, have laws prohibiting its sale and protecting it during its reproductive season. Mangrove forests along Mexico's Pacific Coast contain only about 5 percent of their former iguana population. Guatemala's Pacific lowlands have only remnants left. El Salvador's tropical forests are down to a mere 1 percent of their original density. "I'd be willing to bet that if you plotted iguanas against the size of the human population," says iguana scientist A. Stanley Rand, "you'd find a fairly nice inverse correlation."

Iguana Incubator

In Werner, fresh from earning her Ph.D. with seven years of iguana studies in the Galapagos Islands, the institute had a scientist ready to get her hands dirty in trying to reverse the trend. As a first step, the researcher captured several dozen pregnant females and took them to a natural iguana nesting site 16 miles outside Panama City. There, in sandy soil chosen for ease of egg laying, and within the confines of a 10-by–10-yard pen, she learned her first important iguana lesson: The females excavate such labyrinthine tunnels that the eggs are almost impossible to find.

After digging up the whole area three times, Werner and an assistant finally turned up 700 of the leathery eggs. These were packed into plywood boxes containing dirt, then crammed into a tiny apartment in Panama City, where they were warmed with light bulbs.

With hundreds of baby lizards on hand, Werner quickly designed and built her first iguana ranch on about half an acre at Summit Gardens, a recreational park near Panama City. Anticipating the time when iguanas would be bred by campesinos, she deliberately used materials that can easily be found: roofing metal, bamboo, and palm fronds.

For different research experiments, she constructed enclosures with sheet metal walls to stop snakes wriggling in and lizards clambering out. Wire netting over the top kept out the 'possums that prowled over nightly and the hawks that swooped down to pick off the lizards. Inside were trees for shade and thick branches for sunning, the iguanas' favorite daytime recreation. Also included were tiny compartments—lizards like to squeeze inside tight spaces—constructed from bunches of bamboo, which were raised on stilts and set in trays of water to keep out ants. The lizards' daily fare included fresh-cut leaves, flowers, and such fruits as bananas, oranges, and mangoes.

Most important for the success of the project was Werner's invention of the artificial nest. In nature, gravid females dig tunnels and bury their eggs in a chamber at the end. Werner devised simple tunnels of concrete or clay drainage tubes leading to a chamber of cinder blocks. The blocks form a laying area from which the eggs can be easily collected. "The females are basically lazy," she says. "Give them a tunnel and they'll follow it."

The collected eggs are put into small cylinders made of screening. These incubators, covered with palm fronds for shade, rest in rows on top of the ground, where rains won't drown hatchlings. After the baby lizards break free of their shells, they see only one source of light coming into the nests. They scramble toward it, squeeze through a bamboo pipe, and tumble head-over-heels into a cloth bag attached to the outside of the chamber. There they are collected by staff members, who weigh and measure the creatures, about the length and thickness of a person's little finger, and mark them.

In six months, the captive lizards more than double their length. More than half

reach sexual maturity by the age of two, one year earlier than in the wild. Almost 100 percent survive in captivity, a vast improvement over nature, where 95 percent fall prey to predators in the key years before they reach their second birthday.

Werner's success quickly outstripped the institute's research goals, but last year the fledgling program almost came to a freakish end. When Panama's political climate threatened to dry up iguana research funds, and Costa Rica offered Werner a teaching appointment at its National University, she decided to relocate to Costa Rica. The move became a nightmare.

Within a month, Werner found herself stranded on the Pan American Highway, bound for her new research station with half her entire iguana stock languishing in a truck under the tropical sun. She had made it across the border, where the guards listened incredulously to her description of the truck contents ("Are you crazy or normal?" they'd asked), then had waved her through.

Nature was the real problem. Hurricane Gilbert loosened mudslides across the road, the truck was stuck on a cliff face for hours, and Werner's cargo—1,200 lizards branded and catalogued for their genetic heritage and experimental status—came close to being baked alive before another truck towed them out. "We almost lost five years of research," she says. "The value cannot be expressed."

The Lizard Population Expands

Since her move to Costa Rica, Werner has incorporated both research and implementation under the umbrella of a new foundation to educate villagers and supervise distribution of iguanas to farmers' backyards. So far, 8,000 lizards have gone to repopulation programs in Central America, where their survival rate in natural settings remains high. Werner is working with six communities in Panama and is in the process of involving more in Costa Rica, Honduras, and Guatemala. El Salvador, Nicaragua, Colombia, and Venezuela have also expressed interest in her program.

She has also developed a model for integrating the iguana into existing small farms. The creatures don't require deep jungle, as they like to use the edges of the forest, where they get an optimum amount of sunlight. Farmers can create "shelter belts" along fencerows and easily ranch iguanas along the edges of their farms, helping to safeguard against erosion at the same time. Twenty-one farmers in Panama are now creating such belts. Eventually an improved lizard stock will be developed for farm-based egg laying and hatching right on the patios of local houses.

With some of the 10,000 new hatchlings at Werner's Costa Rican headquarters, she plans to study hereditary and environmental influences on the animals—including social conditions, food, heat regulation, group composition, and sex ratio. "By now we can show it is biologically, economically, and socially feasible to raise iguanas," she says. The next task is to answer hundreds of questions to make the process more effective.

Despite Werner's optimism, the biggest question is whether large numbers of farmers will be willing and able to have their trees and eat their iguanas, too. "It's not such a wacky idea," says the Smithsonian's Ira Rubinoff. "If putting another animal on the table without cutting down forest is not worthwhile, then I don't know what is."

From *International Wildlife,* September/October 1989. Reprinted by permission.

EARTH CARE ACTION

Tree-Saving Stoves

By David Tice

Most efforts to slow global deforestation have focused on planting trees and restricting cutting. But the Bellerive Foundation promotes fuel-efficient stoves as a way

to reduce fuelwood demand. For the price of the average hydroelectric dam in Africa, one could build about 5 million fuel-saving stoves. These stoves, together, would save

Fuel-efficient stoves such as this one at Machokos Teachers College in Kenya can help save tropical forests. (Bellerive Foundation and *New Forests* magazine)

five times the energy the dam could produce.

Bellerive, a nonprofit foundation addressing a wide range of environmental resource problems, modifies stove designs to suit the needs and conditions of specific villages. Stoves range from family-size units to large models for schools, factories, and other institutions. Most are built from clay, steel, and bricks available locally and are often built on the spot.

Bellerive developed a vocational train-ing center at Ruiru, Kenya, to teach technicians to build the stoves and has so far installed more than 400 institutional stoves in Kenya. The Foundation estimates that if it can reach 72 percent of Kenya's 1,750 woodburning institutions, the wood savings alone would amount to 3 percent of that country's forests.

From *American Forests*, July/August, 1989. Reprinted by permission.

Debt-for-Nature Swaps

By Kathryn S. Fuller

No one wins when mounting debt in the developing world leads to environmental degradation. Debt-for-nature swaps, an innovative financial mechanism aimed at leveraging conservation dollars, may be the only international financial transaction in which there are no losers and few risks.

These swaps involve the acquisition of debt by conservation organizations at a discount, its redemption in local currency, and its use for conservation purposes. Such swaps can increase the impact of conservation dollars dramatically. In Ecuador's most recent swap, for example, $1 of acquired debt yielded over $8 worth of local currency for conservation. It is little wonder that conservationists regard this as a unique opportunity not only to protect biological diversity throughout the tropics but to foster sustainable development as well.

The idea of using the debt-equity model to support conservation efforts in developing countries was first proposed in 1984 by Thomas E. Lovejoy, then vice president of the World Wildlife Fund (WWF). International conservation organizations, including WWF, the Nature Conservancy, and Conservation International, as well as local conservation organizations such as Fundación Natura and Fundación Neotropica, pursued the idea vigorously with the financial community. By the end of 1987, swaps had been carried out or were in progress in Bolivia, Costa Rica, Ecuador, and the Philippines.

Bolivia's swap, carried out in July 1987 through Conservation International, involved an agreement by the Bolivian government to set aside 3.7 million acres of tropical forest around the existing Beni Biosphere reserve as a protected area and to establish a $250,000 fund in local currency to manage the reserve. In Costa Rica, a total of $70 million—5 percent of the country's external debts—has been exchanged since 1987 through donation from private groups such as WWF, Conservation International, and the Nature Conservancy, with the bulk of the funds coming from two European governments. The funds have helped expand Costa Rica's park system, create buffer zones around the parks, and establish a major reforestation program. In the Philippines, a $2 million debt swap funded through WWF in collaboration with the Haribon Foundation, a prominent nongovernmental organization based in Manila, has improved management of the national parks and has helped build the infrastructure of the park system.

Making It Work

The Ecuador example highlights some of the key features of successful debt-for-conservation swaps. The concept underlying the Ecuador program originated with Roque Sevilla, then president of Fundación Natura, Ecuador's leading private conservation organization and one of the most dynamic nongovernmental conservation organizations in Latin America. As a result of Fundación Natura's efforts, the Ecuadorean government agreed to exchange up to $10 million of the country's external debt for local currency bonds to be held by Natura. The interest on the bonds—31 percent in the first year—finances a variety of conservation projects, including management of existing and new national parks training for park personnel, and environmental education. At maturation, in nine years, the principal from the bonds will become an endowment for Fundación Natura. In December 1987, WWF used approximately $355,000 to buy $1 million of Ecuador's outstanding commercial debt. Early in 1989, WWF and The Nature Conservancy completed the $10 million program by purchasing $5.4 million and $3.6 million of debt, respectively.

By issuing nine-year bonds rather than cash, Ecuador addressed one drawback to swaps: their potential inflationary effect on the debtor country's economy. In addition, the use of bonds provides a dependable long-term means of support for local conservation. The assured annual revenue from the bonds over the next nine years will allow Fundación Natura to plan and implement long-range programs without the fear of funding shortfalls. Moreover, creating a permanent endowment from the matured bonds for Fundación Natura provides a stable financial base for future conservation efforts. Such stability is critical in developing countries, where the growth of

effective private conservation organizations is more recent than in the United States and Europe.

WWF's recent debt-for-nature swap in Madagascar is yet another success story. Last July, WWF negotiated a three-year, $3-million swap program with Central Bank officials. WWF immediately acquired $2.1 million of the eligible debt, which was converted into local currency. The proceeds will be used in part to train, equip, and support 400 park rangers for Madagascar's priority protected area.

A debt-for-nature swap for Zambia totaling $2.3 million was also recently completed. It is similar in structure to Madagascar's swap. In both cases, proceeds from the swap are in cash and will be disbursed

WHAT YOU CAN DO

Each year the United States imports more than 120 million pounds of fresh and frozen beef from Central American countries. Two-thirds of these countries' rainforests have been cleared to raise cattle, whose stringy, cheap meat is exported to profit the U.S. food industry. Because the beef is not labeled with its country of origin upon entering the United States, there is no way to trace it to its source. Write to the secretary of agriculture and let him know that you want a beef labeling law to specify the country of origin. Write to Clayton Yeutter, Secretary of Agriculture, 14th St. and Independence Ave. SW, Washington, DC 20250.

Rainforest Action Network

over three to four years to mitigate inflationary effects.

With the implementation of six private debt swaps to date, worldwide acceptance for these deals has grown and there are now even more opportunities. With the recent endorsement of debt-for-nature agreements at the Paris Economic Summit, governments are now exploring ways in which they can incorporate this mechanism into their foreign assistance programs. The debt-for-nature concept is ripe for expansion in light of the continuing debt crisis and increasing concern over the state of the global environment.

Cooperation Is Crucial

The conservation community has broken new ground in the use of debt-for-nature swaps. Most crucial to their success, however, is the involvement of local nongovernment and governmental organizations from the early stages of the deal. Debt swaps have little chance of succeeding without a solid relationship between the government and the private agencies in developing countries that are responsible for implementing conservation programs over the long term.

Debt-for-nature swaps offer glimmers of optimism in the struggle to strengthen local conservation programs in developing countries. The mechanism has helped focus public attention in these countries on environmental problems and has led to increased cooperation between the public and private sectors.

Through their involvement with debt-

WHAT YOU CAN DO

Tell the World Bank to stop funding rainforest-killing development projects, such as hydroelectric dams, with your taxes. Dams are costly boondoggles that usually are destroyed within ten years by corrosion and silt. They drown thousands of acres of rainforests, displace indigenous tribes, and saddle developing countries with a permanent mountain of debt, mortgaging their economic futures to U.S. and Japanese banks. Send a letter to the president of the World Bank urging him to stop financing rainforest dams and to fund small-scale projects that benefit rainforests and their inhabitants instead. Write to Barber J. Conable, Jr., President, World Bank, 1818 H St. NW, Washington, DC 20433.

Rainforest Action Network

for-nature swaps, conservationists have developed unprecedented relations with the international financial community. Perhaps other novel ways of funding needed conservation efforts will develop. In the meantime, debt-for-nature swaps represent the best way to stretch the buying power of all those involved in promoting the sustainable use of natural resources. Ecuador's ambassador to the United States, H. E. Mario Ribadeneira, summed it up best when he said of his nation's first debt-for-nature swap, "It is one of those rare cases where everybody wins."

Reprinted with permission from *Environmental Science & Technology*, 1989, 23, 1450–1451. Copyright © 1989 American Chemical Society.

An estimated 100 trillion gallons of water flow down the Mississippi each year, passing through St. Louis, Missouri, on a winding course from Minnesota to the Gulf of Mexico. (Richard Hamilton Smith)

TROUBLED WATERS

By Jerry Adler (with *Newsweek* bureau reports)

Lo, the mighty Mississippi! Most sibilant of river names, from the Algonquin "Father of Waters" (unless it was the Ojibwa "Great Water" or the Chippewa "Big River"). The aorta of North America, stretching 2,344 (U.S. Army Corps of Engineers) or 2,552 (state-park sign at the headwaters) miles from the cold heart of Minnesota to the fingers of the Delta. Like blood, a drop of the 100 trillion gallons that flow down it annually reveals our most secret vices and ailments: the ravages of erosion, telltale ribbons of sewage, alien chemicals bristling with polysyllabic menace. Somewhere along its shores, virtually every environmental problem known to civilization can be seen, except for the destruction of the whales, and if the Mississippi had whales, they'd be in trouble there, too. But the river still lives, even, in some ways, healthier now than when the first Earth Day dawned on its rippling face 20 years ago. The Mississippi! Where else would you go to take the pulse of America?

For the most celebrated river in the nation, the Mississippi can be maddeningly elusive. Around New Orleans, you can drive for miles within a few hundred feet of the river and never catch a glimpse of it behind its levees—which is just as well, because you'd have to look up to see it. On the locked-and-dammed stretch of the upper Mississippi, the water is in plain sight but the river is lost, hidden under the slack lakes and sloughs that have formed in its place. And always, the mud—or, if you please, *sediment*—356,000 tons a day, giving rise to the canard of "too thick to drink and too thin to plow." That is a slander; the Missouri is ten times muddier than the Mississippi.

Its vast volume conceals many sins, some of which are also crimes. The most comprehensive survey of Mississippi River pollution was by Greenpeace USA, which in 1988 found that industries and municipalities along the river discharge "billions of pounds of heavy metals and toxic chemicals into the river every year." That sounds like a lot, and Greenpeace tries to make it sound like even more by including essentially nontoxic substances such as iron,

149

which actually makes up a great bulk of the discharge from some sources. Diluted in the Mississippi's titanic wash, a lethal chemical like dioxin dribbles off into the low parts per trillion. But a trillionth here, a trillionth there—it adds up, when you consider all the pesticides, herbicides, solvents, and trace metals going into the river from the factories, farms and sewers of 31 states. Bob Meade, part of a team of U.S. Geological Survey scientists who have been testing the lower Mississippi since 1987, believes the water is in some ways actually growing cleaner. The big cities on the river now treat their sewage, and lead levels began to decline after most cars switched to unleaded gasoline. But, he says flatly, "we don't eat fish out of the river."

Nor is pollution confined to places downstream of factories and waste-treatment plants. Acid rain does not forbear to fall on the pristine forests of northern Minnesota, where the Mississippi rises from tiny, gin-clear Lake Itasca. "I wouldn't drink this water," says Jack Nelson, assistant manager of Itasca State Park. Of course, he adds, "I'd rather drink it here than in Baton Rouge."

We have, from time to time, made the river over in our own image. In the 1930s, the U.S. Army Corps of Engineers subdued it with concrete, building most of the series of locks and dams that enable barges to climb the 420 feet from Granite City, Illinois, to Minneapolis. On the lower Mississippi, which begins at Cairo, Illinois, where the Ohio in one swoop doubles its volume, the river is too great to be dammed but too dangerous to run free. Here, the corps has hemmed it in with great walls of earth, rocks, and concrete, behind which the people of New Orleans go about their business, as much as 15 feet below sea level. We have gained a lot from the corps's efforts: rich cropland, shelter from devastating floods, valuable industrial sites, and energy-efficient transportation. We are still measuring, though, what we have lost: clean water, irreplaceable wetlands, and the possibility of nature in our very midst. What was once seen as a noble struggle to subdue a wild continent is now viewed by many as a colossal exercise in human arrogance and destruction. It was less than two months ago [February 1990] that Lt. Gen. H. J. Hatch, commander of the Corps of Engineers, advised his troops that "the environmental aspects of all we do must have *equal* standing among other aspects." That is a big change for the corps, which long had the reputation among environmentalists as the part of the Army that makes war on nature. Welcome to the other side, General; with your help, we'll make it, one duck pond at a time.

Red Wing, Minnesota

Everywhere man is crowding the river.

The St. Croix, which forms part of the border between Wisconsin and Minnesota, joins the Mississippi at Prescott, Wisconsin, 20 miles southeast of St. Paul. The two rivers have very different personalities. Downstream from the Twin Cities, the Mississippi has always been a working river, left pretty much to the doleful company of the big barge tows, while the St. Croix for all living memory has been lined with cabins and speckled with the bright hulls of boats. Once every two years, Dan McGuiness of the Minnesota-Wisconsin Boundary Area Commission will photograph the river from the air and count the boats. He has found more than 2,000 boats along his 52-mile route. River patrols enforce a "no wake" rule when congestion exceeds one boat per 10 acres. McGuiness has seen boats crowded one to every two acres. The artificial islands, with their beaches of clean dredged sand, are packed gunwale to gunwale. Latecomers bob in the current, waiting for a spot to open, like cars circling a crowded parking lot.

Increasingly, those boats are venturing out onto the Mississippi. Once, no one who ever flushed a toilet in Minneapolis would have dreamed of boating downstream, but as the Mississippi has grown cleaner, people have begun to use it without getting a gamma globulin shot first. At nearby Lock 3, recreational boats (packed 60 at a time in the 110- by 600-foot chamber) outnumber commercial traffic. Nearly 20,000 pleasure boats were locked through last year, a 70 percent increase in just three years.

Environmentalists, naturally, regard this with alarm. The boats' propellers are said to stir up sediments, cutting off sunlight to aquatic vegetation, which then rots on the bottom, depleting the water of oxygen. The boats' wakes are accused of battering the fragile ecology of the shoreline, upsetting wildlife, causing erosion. But probably the real issue, as framed in an editorial in the Red Wing *Republican Eagle,* is that "each new boat on the river represents a tiny diminishment of Red Wing's quality of life." To accommodate the pleasure-boat fleet, developers want to build some 2,600 new marina slips on the river between St. Paul and Red Wing in addition to the 3,000 already there. Nature lovers fear for the river's dignity as boats named Margie II and Sea What You Done to Me go rollicking from shore to shore in a tinkle of ice cubes.

All the counties on the Wisconsin side of the river have put a moratorium on new marina development while officials study the applications, but it is likely that at least some of them will be built. It is ironic that this workaday stretch of Midwestern scenery should be caught in the same trap as Yosemite, hung by its beauty. Last year [1989] Congress designated the stretch from Hastings north to Dayton as a "National River and Recreation Area." The sponsor, Rep. Bruce Vento of Minnesota, said he would like to see some of the barge terminals and junkyards moved away from the river's edge, to make the water more attractive—for boaters.

Genoa, Wisconsin

For all the vast volumes of water that it carries, the Mississippi is relatively insignificant as a source of hydroelectric power because its terrain is so flat; from Minneapolis to New Orleans it drops only 800 feet in 1,707 miles. It is, however, a vital link in the nation's energy economy. Millions of tons of coal and millions of barrels of oil traverse it in barges. Numerous power plants borrow from the river to cool their condensers, conscientiously repaying it with the same water, 10°F hotter than when it was drawn out.

One such plant is the 350-megawatt generating station in Genoa, Wisconsin, 20 miles south of La Crosse, which in a typical year will burn 800,000 tons—533 bargeloads—of Illinois, Montana, and Wyoming coal. Eighty people work in this plant, producing one-third of the power consumed by 180,000 families in the four surrounding states. Alongside it on the riverbank is a cluster of smaller buildings surrounded by a chain-link fence, topped by barbed wire. Twenty-two people work at this plant, the

La Crosse Boiling Water Reactor, but they produce no electricity at all. The 50-megawatt nuclear plant has been deactivated since April 30, 1987, after its owner, Dairyland Power Cooperative, concluded that it was cheaper to burn coal. In one of the buildings, at the bottom of a pool of filtered and demineralized well water, are 33,300 narrow cylinders, in bundles of 100, containing the spent uranium that once powered the plant. They have become the raison d'etre for the plant. If there were any place else in the world to put them, the reactor could be dismantled—after waiting a few years for its radioactivity to subside—and then everyone here could go home.

The cylinders—8½ feet long and as big around as a fountain pen—contain thousands of eraser-size pellets of uranium, plus trace quantities of its fission by-products: krypton 85, cesium, cobalt, manganese, and plutonium. The uranium was purified from the ore, "enriched" to increase the proportion of U–235 (from a naturally occurring 0.7 percent to 3.7 percent), and sealed in these cylinders, where it continues to fission and generate heat.

The process of creating nuclear fuel doesn't work in reverse. You can't disperse the uranium back into the ore and bury it back in Utah. It is possible to process the fuel for reuse, but what you have left is a high concentration of plutonium. Since the shortcut for anyone seeking to build an atom bomb is to get his hands on plutonium, the government has discouraged commercial users from reprocessing their waste. Back when the Genoa plant was built, it was assumed that the government would dig a hole someplace where no one would care too much, drop in drums of

waste, and cover them up with dirt—just like any other toxic chemical. It hasn't, of course, worked that way. So potent is the fear of nuclear material that the government has been unable for more than 20 years to find a state to accept it. "NIMBY [Not in My Backyard]," says MIT nuclear engineer Kent Hansen, "has become NOPE—Not On Planet Earth."

The latest projection from the Department of Energy is that there may be a working nuclear-storage site in Nevada by the year 2010, but few experts believe it. Even if the dump were to be ready by then, the projected backlog of nuclear waste means that it would probably take at least another ten years before the last of the uranium leaves Genoa. Meanwhile, people have to maintain the pumps that carry away the waste heat, watch the gauges, and patrol the fences, 24 hours a day. For how long? For 10,000 years, give or take, or until the government finds a site for a nuclear-waste dump, whichever comes first.

Cedar Glen, Illinois

About two miles below Keokuk, Iowa, where the Des Moines River comes in from the west, the Mississippi runs past steep, wooded bluffs and keeps Missouri and Illinois apart by barely a quarter mile. This is a historic spot in the domestication of the Mississippi, a few miles below Lock and Dam 19, the oldest of the lock complexes and the only one that includes a sizable hydroelectric generating station. Before the discipline of locks and dams, this was an anarchic stretch known as the Des Moines Rapids. Now the river, running deep and smooth through its narrow channel, conceals its own turbulent history of rocks and ledges and steamboat wrecks, but the countryside is still as wild as any in the well-trodden and picked-over landscape of the Midwest. Three small wooded islands stand just off the Illinois bank, and it is there, in the high branches of the cotton-woods and maples, that the very symbol of American wilderness is making its stand against the depredations of the twentieth century: the bald eagle.

It is an odd concatenation of natural and artificial circumstances that has led the eagles to gather here in the Cedar Glen Wintering Site, one of the largest of the dozen or so official winter sanctuaries in the United States, home to anywhere from 250 to 350 birds in an average season. (The rest of the year, the birds nest in the upper Midwest and Canada.) Less than a mile from the eastern bank is a small, steep-sided bowl whose trees barely reach above its rim, enabling eagles to satisfy their pref-erence for a high perch without having to sit out in the cold wind all night.

When the river ran free, the rapids were one of the few places this far north sure to have open water—and therefore, food—all winter. Now as the water churns through the hydroelectric gates, it picks up just enough turbulence to keep from freez-ing for several miles downstream. And the dam also serves as a fish-processing plant for the eagles, who have a taste, not shared by most humans, for the gizzard shad, a rough, bony fish a little bigger than a big sardine. The eagles prefer shad that are al-ready dead and floating, because those are the easiest to catch. As the fish come down the river past the power plant, quite a few of them are separated from their heads—an ecological success story that Union Elec-tric Co. of St. Louis has oddly neglected to claim credit for.

Of course, the real credit for the eagles' survival, as well as perhaps our own, is due to Rachel Carson, whose *Silent Spring* was the seminal text of the modern environmental movement. Carson, writing in 1962, found evidence that the ubiquitous pesticide DDT was decimating eagle pop-ulations by interfering with their ability to reproduce. It took ten years for the govern-ment to ban the substance, and at the time it was expected to take 20 years for DDT's impact on the environment to dwindle. The bald eagle population in the Lower 48 states bottomed out in 1974 to around 3,000, including only 791 identified nesting pairs.

"This was a disastrous phenomenon in the wildlife kingdom," says zoologist Thomas Dunstan of Western Illinois Uni-versity, who helped establish the Cedar Glen refuge. "A few people kept these birds alive by a thread." Now the nesting pairs have more than tripled, and research-ers estimate there are 8,000 to 12,000 birds—a comeback so impressive that the U.S. Fish and Wildlife Service has pro-posed the bald eagle for one of the highest accolades the government can bestow on an animal, a promotion from "endan-gered" species to "threatened."

Twelve thousand eagles—imagine that! One eagle for every 20,000 Ameri-cans, roughly speaking. No one knows how many eagles the land once supported, but biologists believe it was at least 100,000, which implies that the Indians outnum-bered them by less than 20 to 1. Besides pesticides, civilization brought birdshot, which, even when it missed, filled lakes and rivers—and the eagles' prey—with the

Zoologist Thomas Dunstan found a refuge for eagles in Illinois. (W. McNamee)

insidious poison of lead. Biologists say that eagle hunting is now relatively rare—and severely punished—and many states have outlawed lead shot in favor of steel. But banish all these hazards at a stroke, says Gary Duke, an avian physiologist at the University of Minnesota, and eagles would still face the ravening human desire for habitat. "Eagles like to live on lakeshores and riverfronts, and so does everyone else," says Duke. As the eagles have recovered, they've filled up most of their niche. As one solution, Duke has proposed releasing eagles in the spread-out communities around Lake Minnetonka, near Min-

neapolis. The apotheosis of America: people, crowded out of the cities, and birds, pushed from the wilderness, meeting in the suburbs.

Sauget, Illinois

The southern Illinois river town of Sauget first came to the attention of the state environmental protection agency (EPA) in the late 1970s, when officials heard rumors about a dog that ran into a dry creek bed near the town and emerged in flames. This may or may not have occurred, but the aptly named Dead Creek turned out to contain enough phosphorus to make the soil smolder under the wheels of children's bicycles. The highly combustible phosphorus apparently outweighed the fire-retardant effects of the polychlorinated biphenyls (PCBs) also found there and in half a dozen other spots around Sauget. "You name it, it's there," says Terry Ayers of the Illinois EPA. "These are some of the most grossly contaminated sites in the state."

Based on molecules per capita, Sauget (population, around 200) might be the single most polluted place along the Mississippi. In 1988 the federal government sued the town over discharges from its American Bottoms sewage-treatment plant—more than 16 million gallons of bright yellow-green effluent that goes directly into the Mississippi. "The water was the most toxic we've ever tested," says Dale Bryson, water division director of EPA Region V. This reflects, of course, what goes *into* the sewers. More than half a dozen big chemical and pharmaceutical plants in Sauget and nearby East St. Louis (across the river from St. Louis) discharge their wastes through American Bottoms. The plants are supposed to "pretreat" their effluent, but the government charges that the town has been lax in enforcing this requirement.

It is not surprising that the values of, say, Madison, Wisconsin, are absent in Sauget. The workers at just one Monsanto plant outnumber the townspeople three to one. Sauget in fact *was* Monsanto, Illinois, until 1967, when the chemical company requested a name change to end the confusion over headlines like "Monsanto Man Dies in Fire." The name honors the town's founder and the father of the current mayor, Paul Sauget, who sums up the First Family's credo thusly: "Goddam, they gotta let somebody make a living."

"Just because the government tells you something, it don't mean it's true," Sauget says of the EPA lawsuit. "They try to say we're the dirtiest place on the river. That's just somebody trying to make a name for himself." He fears that the EPA's constantly changing demands have already hurt the town's business. "You don't see smoke anymore, that's what worries me. It just looks like things are slowed up around here."

Here is how pollution happens, and how it can be stopped. Monsanto's plant makes, among some 12 other products, millions of pounds of paradichlorobenzene, the active ingredient in toilet-bowl fresheners and mothballs. The raw materials are benzene, a refinery product that is shipped to the plant by barge, and chlorine, which arrives as a liquid in 90-ton railroad tank cars. They are combined catalytically, and the resulting mixture is purified in several steps, of which one of the last is

to separate the para from its liquid coproduct by crystallization. For years the company used a "wet process." This required curing the cakes of para in a room cooled by thousands of gallons of water, which turned into contaminated waste to be pumped to the sewage plant. Emissions from the para as it cured added the heady odor of mothballs to the variety of interesting smells for which Sauget was and is known. In 1989, Monsanto substituted a "dry process," in which water never comes into contact with the para solution, and the finished product is piped directly to railroad tankers. The company claims this has cut air emissions 99 percent and water emissions from 40,000 pounds a year to zero. This gives the plant a head start on meeting Monsanto chairman Richard Mahoney's self-imposed target of cutting all air emissions 90 percent by 1992 and all land and water discharges by 70 percent by 1995.

Could the same thing be done at plants that make 1-chloro–2-nitrobenzene, 1,1,1-trichloroethane, and all their alarmingly hyphenated ilk? And then what of the finished products themselves—the solvents and pesticides and plastics whose most salient characteristic is that they last for a long time? Pat Costner, the Greenpeace chemist who cowrote the survey of Mississippi River toxins, doesn't think the answer lies in dry versus wet processes but in a decision by society that it can no longer afford to produce certain chemicals at all—in particular, chlorinated hydrocarbons. There are about 14,000 of these compounds in such familiar products as PVC plastics and, yes, paradichlorobenzene.

It was relatively easy to decide to ban

DDT to save the bald eagle. What would you give up to save the moths?

Clarksdale, Mississippi

The eastern shore of the Mississippi, from about the Tennessee/Kentucky border south to Vicksburg, contains some of the richest land and poorest farmers in the United States. This is alluvial land, the distillation of the topsoil of half a continent concentrated on the 50-mile-wide floodplain between the river and the Bluff Hills to the east. When it was growing corn back in Iowa, this soil made farmers rich, but spread over Tunica County, Mississippi, it has led to a 52.9 percent poverty rate, among the highest in the nation. Scattered through it are patches of the dense, heavy clay known as buckshot because it cracks when dry, like a shot-out window; if you're buried in this land, the locals say, when Gabriel sounds his trumpet you'll be just plain out of luck. Water stands on it, making it suitable for rice farming or catfish ponds. The rest is a light, crumbly loam that "melts like sugar and flows like water." Once it supported some of the wildest and richest wilderness in the continental United States—a bottomland forest of oak, elm, and ash and a major wintering ground of waterfowl. Now that it has almost all been cleared, its blessing and its curse is that it is perfectly suited for the heartbreaking, backbreaking, bankbreaking staple of the Delta, cotton.

Cotton is a compilation of bad horticultural traits. It has a raving nitrogen dependency and is prone to rot. The boll weevil is just one of the half-dozen pests it attracts. "Cotton uses just about every pesticide known to man," says Arkansas agronomist Lyle Thompson. That is especially true here, amid the vibrant insect and microbial life of the Delta, where pesticides have long been considered the staff of life for man and plant alike. Malaria and yellow fever have killed a lot more Mississippians than chlorinated hydrocarbons. There are plenty of old farmers who remember washing the walls of their homes with a solution of DDT. "You ban all chemicals tomorrow," says Fred Cooke, an agricultural economist at the Mississippi State University agricultural station in Stoneville, "and in 20 years there'd be no farming and there'd be no towns."

If people ate cotton, they might not be so cavalier about what they put on it, but there are no federal standards for pesticide residue in the crop, and no one (yet) has started charging double in department stores for organically grown T-shirts. The tragedy is what the chemicals have done to the land and the people. Frank Mitchener, one of the largest cotton farmers in Mississippi, facing a bollworm invasion, sprayed his fields with the banned pesticide toxaphene last summer. Nobody might have known but for a 4-inch rainfall that washed most of the pesticide into Cassidy Bayou, killing fish up to 6 miles downstream. Experts say it may take years for the wildlife to recover. The violation cost Mitchener a $55,000 fine. "For the common Joe Farmer out there, it's been like a kick in the gut," says Stoneville agronomist M. Wayne Ebelhar. "They understand why he used the stuff. It's a good, solid chemical that does the job."

And for all that, most farmers aren't even getting rich off their crops. Pesticides are expensive, and with cotton prices not much over 50 cents a pound—down about a third from the peak 20 years ago—the profit margin on an acre of cotton can be thinner than a banker's smile. In a few places along the river, farmers are trying to break out of the cycle of poisoning everything else to grow cotton. A coalition of black farmers around Clarksdale, Mississippi, set out to grow organic vegetables, but they found that in order to escape the countryside's pervasive pesticides, they actually had to locate their plots in vacant lots in town. W. C. Spencer tried to switch his land from cotton to soybeans, the other main Delta crop, but his beans withered and died from the residual herbicides in the soil.

Near Vicksburg, part of a 20,000-acre tract owned by Tara Wildlife Management has been allowed to revert to its native forest cover, which is harvested for high-value hardwoods and rented out for hunting. This is a tiny start toward restoring the 20 million or so acres of wetland that were cleared and drained by man. "The lower Mississippi Valley was to the Temperate Zone of North America what the Amazon is to the Tropical Zone of South America," observes Charles Baxter of the U.S. Fish and Wildlife Service. "It was the incredibly rich ecological system of forested wetlands, rivers, bayous, swamps, and lakes." Only God can make country like this, and even he needs to get a bill through Congress first. Baxter is hopeful that the 1990 farm bill will establish a fund for taking marginal land out of production and restoring it to wetland status, which may actually be cheaper than providing federal crop and flood insurance. "We can't make rivers flood again," Baxter says, "but we can un-

One year, a neighbor's pesticide drove Clarence Hopmann from his Arkansas soybean fields. (J. Ficara)

make drainage ditches." A man with a backhoe can do that in a day, and within the year, wetland plants can be growing in place of soybeans. Someone can drive by with a soybean planter filled with acorns, and in 30 to 50 years, a bottomland forest is reestablished. "For so many years, all we could do was watch the destruction. Now we have the opportunity to turn the clock back a little bit."

In Dumas, Arkansas, just across the river, a farmer named Clarence Hopmann tried to cut down on the use of agricultural chemicals when he developed an allergy to the spray that would drift over from the neighboring cotton fields. But in order to even qualify for loans, bankers demanded that he use large doses of chemicals on his crops. He complied—only to lose his land to the government when soybean prices crashed in the 1980s. He is now trying to reclaim it, full of plans to use the latest organic techniques for rice and winter wheat.

There's only one problem. The government now says that the land he farmed for decades may be actually wetland. He may not be allowed to farm there at all.

Old River Control

Some 50 miles upriver from Baton Rouge, about where the Red River approaches it from the west, the Mississippi begins the slow bend that will take it past New Orleans and, in its unfathomable geologic wisdom, on east to the Gulf along virtually the longest path imaginable. It is here that the Army Corps of Engineers has dared to challenge the full-grown Mississippi and here that the stakes are highest: virtually the entire economy and topography of southeastern Louisiana. If the river wins here, New Orleans is the loser.

A glance at the map shows why. A much more direct route to the sea is followed by the Atchafalaya River, which runs almost due south from that point to empty into the Gulf at Morgan City. In 1831, after the famous steamboat captain Henry M. Shreve cut a channel to eliminate a loop in the Mississippi here, the river seemed to wake up to this possibility. More and more water began flowing down the Atchafalaya. The more that went down it, the deeper its channel became, and the

more water it could draw off. It was easy to predict the outcome if this continued. The Atchafalaya would capture the Mississippi. The former distributary would become the main channel of the river, and the river from north of Baton Rouge to the Gulf would become a slack, saltwater estuary.

New Orleans would survive—at least, the real estate would still be there—but without fresh water and no longer on the main shipping channel, the only industry left would be the one that runs on bourbon. This clearly was a case for Congress. In 1954 it passed a law forbidding the Atchafalaya to take more than 30 percent of the combined flow of the Red and the Mississippi. King Canut should have been so lucky; Congress had the Army Corps of Engineers to enforce its will.

The result is the complex of dams, gates, and channels known as Old River Control, which went into operation in 1963. A major addition was completed in 1986, and the latest improvement—a privately operated hydroelectric station—will begin generating power this summer. The power plant was originally conceived to provide cheap electricity to the town of Vidalia, Louisiana (population: 6,000), but the corps doesn't think small—it insisted it had to be 192 megawatts.

Millions of people rely on the gates of Old River Control. Farmers, fisherman, industry, and the municipalities along both rivers all need water, but in differing amounts at different times, and they are not shy about making their demands known. When the river is low, the crawfish boats in the Atchafalya Basin worry about being stranded and ask for more water. But two years ago a drought reduced the flow past New Orleans to the point that salt-water began infiltrating up the channel, threatening the city's drinking supply. The corps had to tighten the valve on the At-chafalaya to increase the current going down the Mississippi. During the 1983 high water, the state worried that it might have to evacuate the maximum-security Angola state prison, so it asked that more water be diverted at Old River to take the pressure off the Mississippi. That didn't make the soybean farmers along the At-chafalaya happy. Having swallowed the Missouri, the Ohio, and the Arkansas, the lower Mississippi is too powerful to be dammed—but not, it seems, to be domes-ticated and turned into another govern-ment benefits program.

And still the waters lap against the lev-ees and thunder against the concrete rein-forcements at Old River. As land is paved, rivers run higher on the same amount of rain; every parking lot between Ohio and the Rockies is a trough to the lower Missis-sippi. Mark Twain and Raphael Kazmann, professor of civil engineering emeritus at Louisiana State University, are in agree-ment on this point. Man, wrote Twain, "cannot tame that lawless stream, cannot curb it or confine it . . . [or] bar its path with an obstruction which it will not tear down, dance over, and laugh at." Kazmann has made it his crusade to warn Louisiana that a castastrophic failure at Old River is inevitable. Morgan City, he predicts, could be wiped out within two days, despite its 20-foot-high levees. The Corps of Engi-neers disagrees. "We respect opinions," says Bob Fairless, who is a supervisor at Old River, "but the position of the corps is that Old River will operate safely for the foreseeable future."

The "Chemical Corridor"

When the sun sets on the Mississippi at La Place, Louisiana, a scarlet ribbon of smog lights the sky in the west. Between New Orleans and Baton Rouge, great streams of hydrocarbons spew from a hundred plants' smokestacks. Louisiana, which ranks twentieth in population, falls third in an Environmental Protection Agency ranking of states by toxic air emissions. Twenty-five percent of the nation's chemical industry is in Louisiana. Most of it was built along this 150-mile stretch of the Mississippi, which has easy access to the offshore oilfields, plenty of fresh water—and until recently, a state government that didn't look too closely into what Governor Buddy Roemer calls "a deal with the Devil."

The urgent question of the day is whether the Devil's price was just the fa-miliar irritations of smog and ozone, or something more sinister. The Greenpeace report called the area by its local nickname, the "cancer corridor." Compared to the rest of the Mississippi corridor, the ten Louisi-ana parishes where the chemical industry is concentrated "consistently suffered some of the highest rates of mortality from all diseases, [and of] cancer deaths and infant deaths," according to Greenpeace. But the study admits that it cannot prove that these deaths were the result of breathing the air and drinking the water. Many other things could be responsible, such as poverty, high smoking rates, and lifestyle factors. The culture of southeast Louisiana, to put it delicately, does not emphasize the virtues of postponing pleasure.

"Greenpeace committed scientific mal-practice," says Dan Bornáce, president of the Louisiana Chemical Association. Bornáce also asserts that the chemical industry actually returns water to the river "in better shape than when they got it out, in terms of pollution," so he may not be the most open-minded observer. But a disinterested authority, Dr. Joel L. Nitzkin of the Louisiana Office of Public Health, says that "air contamination just doesn't seem to be the answer. If there is a cancer problem related to industry, we believe it is relatively small."

Yes, but what do the people who live there believe? It's not always easy to tell, because the chemical industry keeps buying up the houses around them. The people suspect this is because that's the quickest and cheapest way to settle complaints. But the companies claim this is a responsible tactic to create buffer zones around potentially dangerous chemicals. Georgia Gulf bought out all 40 families who lived in Reveilletown, Iberville Parish, and settled a class-action lawsuit over alleged pollution. Dow Chemical is trying to buy the community of Morrisonville, near Baton Rouge. In 1982, after accidental emissions and 78 fires at the now defunct Good Hope Refinery and numerous evacuations of the next-door elementary school, the oil company bought the school.

Among those who remain, suspicion is rampant. In 1987, Kay Gaudet, a pharmacist, reported to the state an unusually high rate of miscarriages in St. Gabriel. But a study by Luann White, associate dean of public health at Tulane, found that miscarriage rates in the parish were about the same as elsewhere in Louisiana. Oddly enough, this reassuring news was not well received in the community. "They were not really happy to hear this, because they were sure there had been an increase [in miscarriages] due to environmental problems." White says. "When you're dealing with the environment, there are emotional concerns, political concerns. Sometimes the concern taken least into account is science."

Terrebonne Parish

Past New Orleans, the Mississippi rushes on for another 115 miles to its triple outlet on the Gulf. Between here and the mouth of the Atchafalaya, 110 miles to the west, lie the Cajun swamp parishes of Lafourche and Terrebonne. Here the land doesn't so much meet the sea as gradually shred into it. Cypress stand in clumps of real estate surrounded by water, brushing the ripples with their trails of Spanish moss. Inexorably the real estate is shrinking. Louisiana wetlands are disappearing at the rate of 50 square miles a year. An aerial photograph of Terrebonne Parish from, say, 30 years ago, would show a solid green carpet of marsh. Today, says Oliver Houck, professor of environmental law at Tulane University, "you see something that looks like a torn rag."

This is yet another environmental catastrophe perpetrated, in innocent pursuit of its congressionally mandated goals, by the Army Corps of Engineers. In its natural state, Terrebonne Parish was a shelf of silt just above sea level, continually compacting and sinking but eternally renewed by the sediment in the floods that washed over it every year. Now, as the river runs to the Gulf between its formidable levees, the sinking goes on, but there is nothing to rebuild with. "The Mississippi over the course of history wriggled across Louisiana like a hose, spraying all this life-giving stuff," says Houck. "When we decided to dedicate the river to navigation, we altered the balance. We have taken the hose and turned it off"—or, more precisely, directed it straight down a sewer, in the interest of efficiency.

Of course, the levees are not wholly to blame; how could there be an environmental catastrophe in Louisiana without the oil industry? To clear a route for drilling rigs, oil companies and the corps cut more than 10,000 miles of channels through the Louisiana marshes and swamps. These typically are built 100 feet or wider, but the destruction doesn't stop there; after the interlocking root systems of the marsh are severed, the canal banks may erode to three times their original width. This makes for a unique kind of wetland loss. Elsewhere on the North American coast, the shore is being washed back; here, the marsh is falling apart from within.

Ironically, the loss of swampland is accelerating just as its most characteristic inhabitant, the American alligator, is making a comeback. Hunting, banned for years, was reinstated in the early 1980s. Actually, most alligators are trapped on baited hooks left in the water overnight; that way they can be taken without damaging the skin, which has recently been bringing up to $58 a foot. "Alligator Annie" Miller, a tour-boat operator in Houma, now goes out just before hunting season with a boatload of chicken parts for her favorite animals—the ones who will come to the surface and entertain the tourists when called. Then they won't feel like feeding until the hunters have their limit of one alligator per 150 acres.

Louisiana, a poor, rural, and conserv-ative state, was slow to awaken to the ecological imperative. It is still very tough to take on the oil industry in the state legislature, but by referendum last year the people voted (by a 75 to 25 margin) to set aside a portion of oil and gas severance taxes to wetlands restoration. In line with President George Bush's declared national policy of "no net loss" of wetlands, Gov. Roemer has set the same goal for the state to achieve by 1992, although the oil-tax revenue, a maximum of $25 million a year, seems inadequate to the job. All over the country people are uniting to save wetlands, one shopping center at a time, but 80 percent of American's wetland loss is occurring wholesale right here, home to nearly a third of the nation's seafood harvest. "If this were happening on Chesapeake Bay, it would be getting daily attention," says Bob Jones, the Terrebonne Parish engineer. "If we were losing 40 to 50 square miles a year to the Cubans, you know what kind of money we'd be spending on it."

The Gulf

With momentum gathered over 2,000 miles, the Mississippi falls off the Delta into the Gulf, and the currents push it east to west, making a green-brown swirl in the blue waters. A continent's worth of nutrients drop into the sea and commence to make their way up the food chain in the direction of blackened redfish. Fishermen, shrimpers, and oystermen all gather here for the wriggling bounty of the sea.

But sometimes what the river brings is death. Along the fishing villages of Mobile Bay, 130 miles from the Delta, there is a rare phenomenon the locals call jubilee. It hap-

WHAT YOU CAN DO

When changing your car's oil, anti-freeze, or other fluids, don't pour the old material onto the ground. It will leach out into the groundwater or into streams, rivers, and oceans. It takes only a little oil to cause a lot of damage—1 quart of oil can contaminate up to 2 million gallons of drinking water. Take used oil and other automotive fluids to local service stations for proper disposal. Many states have established oil recycling centers; find them and use them.

Center for Marine Conservation

pens in the late summer months when the air is heavy and still and the Gulf waters are as warm as blood. Then swarms of shrimp, crab, and deep-sea fish may suddenly appear in the shallow water just off the beaches, swimming suicidally toward the shore—and the waiting nets, not to mention the pots, of the lucky towns-people.

To biologists, these are not fish that are seeking to get eaten but fish desperately searching for oxygen. The Mississippi, in its ceaseless fertilization of the continental shelf, can contribute to a condition known as hypoxia. Nitrogen from the rivers produces a bumper crop of algae, which sink to the ocean floor and decay, using up oxygen. When conditions are right, the fresh river water can settle like a blanket over the heavier saltwater, trapping fish in their asphyxiating bath. Farther from shore, scientists have found "extensive, severe, and long-lasting" hypoxic zones almost every

year since 1985, covering areas up to 4,000 square miles, according to researcher Nancy Rabalias. Environmentalists, with their natural bent for drama, call this a dead zone.

This may well be an entirely natural phenomenon. Or it may not; some evidence suggests that the area affected by dead zones is growing by as much as 19 percent a year. The "pregnant question," according to Donald Boesch, director of the Louisiana Universities Marine Consortium, "is whether man is doing something to intensify it." The nitrogen load of the river has been increasing more or less steadily since the 1950s. Nitrogen compounds are a ubiquitous industrial pollutant. Fertilizer runoff and sewage outflow are major sources. Burning fuel in cars and power plants creates nitrogen oxides, which precipitate out as one form of acid rain. A similar suffocation-by-algae has afflicted lakes, sometimes killing virtually everything in them. But this is the sea, swept by currents that dwarf even the mighty Mississippi. Could the same thing happen to *the entire Gulf of Mexico*? Are we seeing "a major change in our natural history," as R. Eugene Turner, chairman of marine sciences at Louisiana State University, puts it? Or is Texas A&M oceanographer Robert Presley right when he says, "I simply do not believe these horror stories that the Gulf of Mexico is being destroyed by man. It's just plain physically impossible."

The fishermen, for their part, go out with their nets and do their job; if the catch is good, they thank God and the Mississippi, and if it's not, well, they'll have better luck another day. Here the river lies beyond the reach of the Corps of Engineers and the Minnesota-Wisconsin Boundary Area Commission and the yellow torrents of sewage. The rest of us have only the river we have made—a river that flows on command and runs where we tell it, whose water comes to us processed through the gates of 29 dams, every drop of it tainted and corrupted by civilization before being allowed to seek its solace in the sea. It carries with it the sorrows of a continent.

 EARTH CARE ACTION

Once Aflame and Filthy, a River Shows Signs of Life

By Bill Lawren

Just before noon on June 22, 1969, Ohio's Cuyahoga River burst into flames. A burning slick of kerosene and oil floated downstream through the city of Cleveland, issuing five-story flames that all but incinerated two railroad bridges spanning the river. And on the heels of the flames came fires of ridicule.

Cleveland's burning symbol of urban filth shocked Americans into a new understanding of the nation's pollution crisis and lit a fire in the hearts of environmental reformers marching toward Earth Day. But the city with a flammable river became the butt of jokes on radio and television. And long after the fire was out, the river continued to be a source of keen embarrassment to city officials.

Those officials blamed waterfront industries for profligate dumping of wastes that had transformed a once-sparkling stream into a sewer virtually devoid of life. A few carp and sucker fish still survived at the very bottom of the river, but even the mayflies on which fish fed had been driven away by toxic chemicals.

Stung by derisive publicity, city and state officials passed regulations limiting industries from discharging toxic metals, acids, and other chemicals into public sewers. The Cleveland Sewer District upgraded its treatment facilities until they are now, according to Water Pollution Control commissioner Darnell Brown, "among the most extensive in North America."

The result? Twenty years later, Cleveland's infamous flaming river appears on the mend. The water again sparkles on sunny days, and pleasure boats, dockside taverns, and upscale condos line the banks. Equally important, biologists who monitor the Cuyahoga's vital signs say mayflies have returned in droves, and with them as many as 27 species of freshwater fish. On summer weekends, the dock of Settler's Landing is crowded with kids dangling fishing lines into what was once wastewater.

But no one maintains the battle to save the river is won. Industry is fighting even stricter water-pollution control standards that the Environmental Protection Agency wants to impose, and sewage still overflows into the river during heavy rains. There are no jokes about the Cuyahoga on late night TV these days (perhaps because other environmental disasters long ago stole the spotlight), but city officials concede that their recovering river remains ill. In fact, the kids fishing at Settler's Landing might want to chew over these words from Pollution Control Commissioner Brown: "The number of fish with tumors has slackened considerably, but I still wouldn't entirely endorse eating them."

From *National Wildlife*, February/March 1990. Reprinted by permission.

EARTH CARE ACTION

Saving a Bay with Curbside Warnings

By Gelareh Asayesh

If you're going to drop junk down a storm drain, don't be surprised if it pops up next to your boat on the bay. Better still, just don't do it.

That's the message Prince George's County, Maryland, is trying to get across to its citizens with stickers at every storm gutter in the county warning: "Do Not Litter— Chesapeake Bay Drainage."

County officials hope the program— which is to include all 32,000 gutters in Prince George's in the next two years—will deter people from dumping waste, ranging from motor oil to grass cuttings, into storm gutters that carry the material straight into the bay.

Prince George's is the first county in the state to place warnings on its storm gutters, although other localities encourage community groups to do so.

"Storm drains are for rain only and not for oil and not for grass clippings and not

Working with stencils provided by the Chesapeake Bay Foundation, student volunteers paint an important message over storm drains. (David Harp/Chesapeake Bay Drainage)

for Styrofoam," said County Executive Parris N. Glendening, standing before a freshly painted and stickered gutter in front of the county administration building in Upper Marlboro.

"Some home mechanics and home gardeners dump their oil waste and yard waste in storm drains on Saturday. On Sunday they go to the bay and bemoan the oil slicks and other pollutants," Glendening said. "Some people think it goes to sewage treatment plants, but not so."

Restoring the Balance

County environmental officials estimate that about half of the organic pollutants in the bay—nitrogen and phosphorus—come from storm water runoff collected by storm gutters.

These elements, when present in extreme quantities, trigger organic decay that deoxygenates water and has disrupted the bay's ecological balance.

Though much of the problem is caused by storm runoff, illegal dumping in storm gutters is a significant part of the problem, said Thomas W. Burke, Jr., director of Governor William Donald Schaefer's Chesapeake Bay Communication Office.

Because the dumping is done surreptitiously, environmental officials cannot estimate the extent of damage, he said.

"This is critical for the health of the bay," said Burke, who presented a congratulatory proclamation to Glendening from Governor Schaefer. "It gets people to make the link that [water from] the storm drain winds up in the bay. No one ever told them that before. It's just like you walked up to the Bay Bridge and dumped your garbage in the bay."

"This is consciousness-raising," said Victoria L. Greenfield of the country's Department of Environmental Resources. County employees hope to place the stickers on about 2,700 gutters this year, starting with 96 drains in the Beacon Heights community.

Warnings on storm gutters first turned up in Maryland in Anne Arundel County, which in 1987 began distributing paint and stencils to interested scout troops and other community groups.

Since then, groups in about 13 communities have taken advantage of the program, and county planners have received queries from around the country, said Patricia J. Haddon, a county environmental planner. Anne Arundel's signs are in green paint and read: "Don't Dump, Chesapeake Bay Drainage."

The Annapolis-based Chesapeake Bay Foundation also has started a similar program. Just this summer [1989], the foundation issued a brochure describing the program, which offers stencils and information to interested schools. Montgomery County will kick off a similar plan next month [September 1989], as will Baltimore.

From the *Baltimore Sun*, August 24, 1989. copyright © 1989 The Baltimore Sun Company. Reprinted by permission.

 EARTH CARE ACTION

Dinosaur Dam

By Sidney Atkins

The Nagara River, which flows out of the mountains of Gifu and into Ise Bay, is a very special waterway. Famous with tourists for torch-lit cormorant fishing on summer nights, the Nagara is now attracting the urgent attention and concern of environmentalists nationwide. It is one of only two remaining undammed rivers in all Japan and the only one on the main island of Honshu. The Nagara River's ecosystem is relatively intact. Its waters teem with fish, including migrating sweetfish and a unique warm-water salmon, which lives only in the Nagara. These are wild fish, not the bland hatchery products that have supplanted the original species in most of Japan's impoverished and abused rivers.

Despite a fairly dense population settled in villages, towns, and cities along its banks, the Nagara River is still fairly clean, especially so in its upper reaches where one can safely drink the water and support a family with one fishing rod.

Incredibly, the Ministry of Construction, the Public Corporation for Water Resources Development, and local politicians are trying to destroy this near-ideal relationship of people and natural environment by building an enormous 661-meter-long tidal barrier that will shut out the sea, blocking the exchange of saltwater and fresh water, and thus prevent fish from migrating. At the same time, the dam and large-scale dredging upstream will create a reservoir that will destroy the rich tidal zone and brackish water habitat of clams, crayfish, loaches, and other creatures. Construction has already begun and is expected to be completed in five years.

This is a dinosaur development project left over from the 1950s that will cost the public millions. The building of the dam and the 30-kilometer-long reservoir is a sop to the construction industry, the major source of most politicians' campaign funds.

But great crises can also be great opportunities. Until last year, the Nagara River's fate was a local issue, fought over in Japan's cumbersome courts and negotiated between the public development corporation and local fishermen's unions. Then in 1988, just when legal means of stopping

the dam were exhausted and construction began, the movement to save the river was resurrected in a new form and has broadened into national debate on the future of all Japan's rivers. Much of the credit for this revival must go to Reiko Amano, an Osaka-based writer and the first female in Japan to be publicly recognized as a fishing master. Her appeals to save Japan's last healthy, free-flowing river led to formation of a broad-based coalition of fishermen, canoeists, biologists, writers, photographers, academics, and local residents. These people who knew and loved the Nagara took out newspaper advertisements opposing the dam's construction. Opposition groups mushroomed in Gifu, Nagoya, Osaka, and Tokyo. Famed author and fisherman Ken Kaiko lent a hand as formal spokesman for the movement.

As Japan's last truly living river, it must be considered a natural legacy belonging to all—if only as a sample of what all the rivers once were and could be again. The Nagara River's supporters are heartened by suspension last year of a similar giant estuary dam project at Lake Shinji in Shimana Prefecture after it had been 80 percent completed.

From *Resurgence*, November/December 1989. Reprinted by permission.

The Courage to Speak Out

By Mark Van Putten

Long before the first Earth Day celebrations took place, Fred Brown was in the front lines of his home state's environmental battles. As president in 1969 of the Michigan United Conservation Clubs (MUCC)—a National Wildlife Federation affiliate—he led the organization into a court fight over a plan to allow industry to impound state rivers and sell "surplus" water. The fact that the Dow Chemical Co., Brown's employer at the time, had aggressively promoted the plan never tempered

Scientist and environmental activist Frederick Brown doesn't shrink from the paperwork involved in protecting Michigan's waterways. (Kevin Horan)

his determination to do the right thing. The court eventually found the plan unconstitutional.

Now, at Earth Day year 20, Brown is still doing the right thing when it comes to issues of water pollution and public access to water in the Wolverine State. He has served an unprecedented three years as volunteer chair of Michigan's Water Resources Commission, which oversees all of the state's water pollution control programs. He spends countless hours meticulously examining every pollution discharge permit request. He also serves on several Environmental Protection Agency advisory committees concerning water quality in the Great Lakes. And for ten years, he was a member of the Michigan Environmental Review Board, evaluating the potential impacts of proposed state actions.

Clearly, his record of environmental activism is impressive. But there's more. Brown's 28-year career with Dow as a researcher makes him a remarkable example of courage and conviction. While working for the huge Midland-based chemical complex, he researched and developed new products for the forest industry. At the same time, though, he used his scientific expertise to minimize water pollution created by Dow, the paper industry, and countless other companies.

In 1979, without legal help, Brown

WHAT YOU CAN DO

If you must use harsh chemicals, don't pour them down the drain or into storm sewers. When you are ready to dispose of them, keep them in their original containers, tightly sealed and wrapped, and put them with your regular trash. Many communities have special collection days for such materials; find out if yours does.

Center for Marine Conservation

says the 65-year-old scientist, who retired from the company in 1982.

Whether Brown's career at Dow suffered from his activism is difficult to say. What is clear is that he never allowed the prospect to stifle his activism. "Dow was more afraid of Fred than he was of them," says MUCC executive director Tom Washington.

In 1984, Michigan governor James J. Blanchard appointed Brown to the "environmentalist" seat on the Water Resources Commission, a position that has become an unsalaried "full-time job" for the hard-working conservationist. Governor Blanchard, who has since reappointed Brown to the position, calls him "Michigan's citizen watchdog of water quality." And indeed, the people of the state owe a debt of gratitude to a man who has spent most of his adult life speaking out for their environment, no matter what the potential personal costs.

filed a challenge to the wastewater discharge permit for the Midland Nuclear Power Plant, which was planned to supply electricity and steam to Dow. A few years later, he relentlessly pressed the Water Resources Commission to impose the toughest toxic pollution controls in the country on Dow. "I simply applied what I learned at Dow to protecting the environment,"

From *National Wildlife,* October/November 1990. Reprinted by permission.

She Slew Goliath

By David Seideman

From the moment New York City planners conceived the Westway Project in the late 1960s, the design seemed destined to be executed. The multi-billion-dollar plan called for the construction of a 12-lane highway that would traverse and tunnel through more than 200 acres of landfill in the Hudson River. It was supported by an impressive list of luminaries, including a U.S. president, two U.S. senators, two governors, and a mayor. None of them, however, bargained for having to deal with Marcy Benstock, a budding Manhattan activist who clearly was against the development scheme.

"These politicians, if they ever see the Hudson River at all, only see it as a vacant lot," says the 46-year-old Manhattan resident. In what may be the most effective grass-roots campaign in New York City history, Benstock mobilized thousands of volunteers, testified at endless hearings, lobbied legislators, and pored through reams of documents to find holes in the Westway supporters' case. Among her coups was her publicizing of government studies that proved that planners had failed to understand the project's potential impact on young striped bass. More than half of the Eastern Seaboard's entire population of the species spawn in the Hudson.

Finally, after 15 years and $200 million

in preliminary state and federal expenditures on Westway, Benstock and her allies helped convince a federal district judge and the U.S. Congress that the project was a threat to the river's integrity and a waste of energy. Westway was scrapped in 1985. "I hope next time we will be on the same side," wrote New York governor Mario Cuomo in a congratulatory telegram to Benstock.

Today, however, Benstock's battle to safeguard the Hudson from destructive development is far from over. City and state officials have repeatedly proposed other plans that could threaten the river's ecosystems—plans that Benstock has usually spoken out against. "What we're talking about is preventing the paving over of a critical habitat," she says. "I want to see the wise use of our limited resources, sensible uses of public funds, and environmental protection."

It's a refrain Benstock has echoed since the early 1970s, when she became concerned about the poor quality of the air she was breathing in her Upper West Side neighborhood. She organized citizens into a group known as Clean Air Campaign, which she still leads today. Within two years, the group had helped reduce the level of airborne soot in the neighborhood by more than one-third. "If you looked up

Can't fight city hall? Marcy Benstock did—and won, blocking a plan that would have upset the ecology of the Hudson River. (Dana Tynan)

the word *tenacity* in a dictionary, you'd find Marcy's picture," says Eric Goldstein, an attorney with the Natural Resources Defense Council.

Adds Manhattan borough president Ruth Messinger, "I've known Marcy since the earliest days of the Westway Project, and I know she's not one to give up. The result is that she has won many victories and probably will win many more." Benstock's record undoubtedly spells trouble for developers with big eyes on the Hudson.

From *National Wildlife,* October/November 1990. Reprinted by permission.

Healing a Sick Bay

By Kathryn Phillips

Tacked amid a collection of colorful environmental posters on a wall in Dorothy Green's office is a decidedly uncolorful chart—a construction schedule for upgrading the main sewage disposal plant for the city of Los Angeles. For Green, the chart is a constant reminder that you can successfully fight city hall—and potty train it, too.

Green's fight began in early 1985, when she and a handful of concerned L.A. citizens learned the city was trying to get a waiver from the Environmental Protection Agency (EPA) to allow it to continue dumping inadequately treated sewage into the Santa Monica Bay. By midsummer, a new organization called Heal the Bay had been born in Green's West Los Angeles living room. The goal: to stop sewage contamination of the 175-square-mile bay and its adjoining wetlands.

For nearly 40 years, the city of Los Angeles has dumped as much as 420 million gallons of sewage each day into the Pacific from its Goliath Hyperion plant, the city's largest treatment plant on the Santa Monica Bay. The 1972 federal Clean Water Act and state laws required the city to provide full secondary (and thus complete) treatment before the end of the 1970s for all sewage. But the Hyperion plant continued to provide such treatment for only 25 percent of its load. And rather than comply with the laws, L.A. officials spent 13 years seeking waivers. Meanwhile, evidence mounted that fish and other marine life were being contaminated and even wiped out by various toxic substances carried in sludge and sewage.

With Green as president, Heal the Bay took its cause to the beaches of Santa Monica, holding rallies to gain public support. It also played an active role in an EPA lawsuit that forced the city in 1986 to stop dumping 4.5 million gallons of sludge a day from the Hyperion plant into the bay. And it took part in negotiations resulting in a $4 billion commitment by the city to overhaul its treatment system.

Green's style of activism is "very cooperative. She's not all confrontational," says Catherine Tyrrell, director of the Santa

WHAT YOU CAN DO

Fill a gallon plastic bottle with water and place it in your toilet tank. (A brick will dissolve and clog the sewer.) You can save up to 5,000 gallons of water a year. Or consider replacing your standard toilet with a water-saving model, which can use up to 70 percent less water. More and more models are now available.

Center for Marine Conservation

Monica Bay Restoration Project, a joint state and federal program.

Today, Los Angeles is in the process of building a new secondary treatment facility and raw sewage overflows into the bay, once routine and unreported, are now rare and highly publicized. The city also has begun recycling sludge and reclaiming water. "Our main accomplishment," observes Green, "has been just to turn city hall totally around on this."

In recent months, Heal the Bay's membership has grown to 8,000. The group has hundreds of active volunteers—from movie stars who help raise funds to lifeguards who help scout for runoff pollution problems. "It's an avenue," says Green, 60, "for people to get involved in protecting their environment."

From *National Wildlife*, October/November 1990. Reprinted by permission.

Standing 5 feet tall (North America's tallest bird), with wings spanning 7 feet, the majestic whooping crane certainly has reason to whoop: It is headed back from the brink of extinction. See story on page 183. (Frank Oberle)

CAN THEY BE SAVED?

By William F. Allman, with Joannie M. Schrof

Six tons of regal beauty, an elephant glides across the savanna with a seeming indifference to the ways of man. Yet the beauty of the elephant's ivory tusks is also its curse, and in man's quest for "white gold" he has been anything but indifferent.

Lured by the opportunity of garnering a year's wages with a brief squeeze of the trigger, poachers have slaughtered elephants by the tens of thousands to supply a booming international trade in illegal ivory. A decade ago, Africa's elephant population stood at more than a million; it has now been slashed in half. Scientists predict that if the carnage continues, not only will the giant beasts disappear but so will the habitats that are home to dozens of other species of animals. With their voracious appetite for trees and other vegetation, elephants turn jungle into savanna, opening up grassland for grazers such as antelopes and wildebeests.

Shocked by the extent of the African elephant's decline, conservationists around the world are launching an all-out campaign to save the animal from extinction.

In Kenya, whose elephant population has been reduced by 70 percent over the past ten years, game wardens have been issued new vehicles and automatic weapons, are backed up by surveillance aircraft—and have been given new orders to shoot poachers on sight. There are signs that the emergency measures are taking effect: Since June [1989], the rate of elephant killings has been cut from three per day to about one a month, and more than 20 poachers have been killed.

Crucial to saving Africa's elephants, however, is stopping the ivory market at its source—the ivory shops from Asia to the United States. An important first step in this effort may come at this month's [October 1989] meeting in Switzerland of the Convention on International Trade in Endangered Species, known as CITES. Prompted by proposals from Kenya, Tanzania, and several other African nations and supported by conservation groups around the world, CITES is expected to agree to a ban on ivory sales in the United States, the nations of Europe, and many

175

other countries. Japan, whose use of ivory in jewelry and personalized signature seals, called *hanko,* makes it one of the world's largest ivory consumers, will at least make a commitment to tighter controls on its imports. South Africa, Zimbabwe, and Botswana, where elephant poaching is not widespread, have opposed the ban because they sell ivory legally to help support their parks. But conservationists insist that any legal trade in ivory provides a conduit for illegally gained tusks; an estimated 70 percent of the carved ivory that is legally sold today actually originates from illegal poaching.

Supporters of the ivory ban argue that, unlike the CITES prohibition against the trade of rhinoceros horn 14 years ago, which did little to stem rhino poaching, this ban will work. In powdered form, rhinoceros horn remains highly valued as a medicine in Asia, they point out, but ivory was long ago replaced by plastic in such products as billiard balls and most piano keys. "The only reason elephants are dying now," says Richard Leakey, the renowned paleoanthropologist, who was recently named director of Kenya's Department of Wildlife Services, "is that Tom, Dick, and Harry want to wear baubles and trinkets made of ivory." Conservationists hope that the outcry over the slaughter of elephants, and an international prohibition to back it up, will make wearing ivory taboo.

Leakey and other conservationists realize, however, that even if the ivory trade is stopped, their battle for the elephant's long-term future is just beginning. "We have to recognize that the days when there were large tracts of Africa for animals to roam freely are gone," says Leakey. "Those days are gone in Europe, they are gone in the United States, and they are gone in Kenya." If elephants are to survive into the next century, conservationists know that they must balance the elephant's appetite for more than 300 pounds of food a day and migration patterns that cover hundreds of miles with Africa's soaring population growth, scarce farmland, and desires for economic development. Even the Asian elephant, whose tusks are typically too small to attract poachers, is in danger of extinction because of its rapidly disappearing habitat.

F rom tropical rainforest to frozen tundra, the ever-expanding presence of humans is colliding head-on with wildlife around the world.

The importance of finding a balance between the needs of humans and the needs of wildlife is not confined to elephants of Africa. From tropical rainforest to frozen tundra, the ever-expanding presence of humans is colliding head-on with wildlife around the world. Environmentalists in the United States trying to save the flora and fauna of the ancient forests in the Pacific Northwest are battling local loggers who see the trees as a means to feed their families. Each year, thousands of acres of Amazon forest are burned to provide land for Brazil's ranchers and subsistence farmers, eliminating hundreds of species of plants and animals in the process. In the environmental outcry over plans to drill for oil in an Alaskan wildlife preserve, a fundamental issue is often obscured: As long as the people in developed nations want the lifestyle born of oil, they will have to drill into somebody's land to get it, most likely threatening the existence of some species.

A dramatic example of the ongoing clash between wildlife and human needs is Kenya's Amboseli Park. Nestled under the towering visage of Mount Kilimanjaro near Tanzania, Amboseli's exotic menagerie of zebras, antelopes, and elephants is one of Kenya's most popular tourist attractions.

The park is not as popular with the members of the Masai tribe who live nearby, however. When Amboseli Park was created 15 years ago, the Masai were pushed off the 150-square-mile parcel of land whose streams had supplied most of the water for their cattle. In return, the ranchers were promised a series of watering holes built by the government for their cattle.

In the competition for resources between the wildlife and the local ranchers, however, the relations between people and park began to sour. The water system broke down soon after it was built and was never repaired, so the Masai illegally began driving their cattle into the park for water and grazing, threatening the park's fragile ecosystem. Friction increased when some Masai began to grow crops on land nearby, which were occasionally trampled by wildlife wandering outside the park's boundaries. Though the Masai have long had a reputation for peaceful coexistence with wildlife, they began to harass the animals that wandered near their crops. More problematic, they turned a blind eye to poachers who killed elephants that had migrated out of the park.

As a result of the struggle between Amboseli and its Masai neighbors, the park's entire ecosystem is now in jeopardy.

As poaching increased, the elephants, who normally migrate freely, began crowding into the safer confines of the park. The great beasts' foraging habits, which usually give rise to the mixture of trees and grass typical of savannas, instead led to widespread deforestation. What was once a park covered by dense thickets of acacia trees is now mostly barren grassland supporting only a few shrubs. As a result, the elephants are putting further pressure on the park's remaining vegetation and in the end threatening their own existence as well.

Over the next century, the human population is expected to increase by roughly 5 billion, mostly in developing countries that, not coincidentally, are also the last refuges for wildlife on the planet.

David Western, a Kenyan biologist with Wildlife Conservation International, discovered just how devastating elephant overpopulation can be when he built an electric fence around his home on the park grounds. Erected eight years ago, the fence consists of a band of wires suspended high enough to keep elephants out while allowing other wildlife, such as wildebeests and zebras, in. Outside the fence is sparse grassland; inside the fence is a thick stand of acacia trees, some of them 20 feet high.

Around the world, the competition between wildlife and people typified by Amboseli will only become more intense in the future. Over the next century, the human population is expected to increase by roughly 5 billion, mostly in developing countries that, not coincidentally, are also the last refuges for wildlife on the planet.

Biologists warn that the destruction of wildlife habitat as a result of the rapidly expanding human population is causing the extinction of species of plants and animals at a rate unprecedented in human history. According to biologist Edward O. Wilson of Harvard University, as many as 6,000 species are going extinct each year from deforestation alone, a rate 10,000 times greater than before the appearance of humans on the planet.

Biologists warn that with the widespread destruction of the world's plant and animal species we are also losing a valuable source of potential new foods, drugs, and genetic diversity. They point to the rosy periwinkle, a tiny plant that contains natural chemicals that have proved effective against Hodgkin's disease, and a rare variety of hardy corn possessing genes that enable it to grow perennially, a unique trait that might be conferred on other varieties of corn through genetic engineering.

In trying to rescue wildlife from the squeeze of human expansion, the lesson of Amboseli is clear: Appeals to the grace and beauty of an elephant—or any other species—are just not enough. A farmer may admire an elephant from afar, but when it is trampling his crops, the beast is a threat to his livelihood. In a world of scarce resources where many people lack food and adequate health care, the question of how to save the elephant or other species from extinction inevitably leads to a larger and more pragmatic question: *Why* save it?

Leakey and his allies believe they have an answer: Saving wildlife is a choice not only of the heart but of the pocketbook. For Leakey, the elephant represents Africa's heritage, but it also provides a means for Africa's people to meet their growing economic needs.

It is a view that is increasingly finding favor among conservationists. For years, environmentalists have been content to save a parcel of wilderness by merely fencing it in and leaving it alone. But in countries where people live day-to-day off the land, ignoring the needs of local people living near the park is naive and risky. "Many people are desperately trying to make it through the next 48 hours," says one conservationist. "If the life of my family were at stake, I'd probably shoot an elephant, too."

The most compelling strategy for saving wildlife, many conservationists now argue, is to ensure that it is economically beneficial to the people who must coexist with it. Mixing pragmatism with conservation is the keystone of a new environmental strategy called "sustainable" development. Conservationists argue that prudent, long-term use of wild lands is far more environmentally sound—and in the long run more profitable—than short-term exploitation. The soil beneath a patch of burned rainforest, for example, is so impoverished of nutrients that it will yield crops for only a few years. But selective logging of the forest can be carried on indefinitely. And in a recent paper in the British scientific journal *Nature,* researchers calculated that the economic value of harvesting the fruits, oils, rubber, and cocoa in a standing rainforest is nearly two times greater than the forest's value if it were cut down and used for lumber or for grazing.

Better Than Coffee

A more direct marriage of environmental concern and economic development is "eco-tourism." Lured by the chance for an exotic experience, tourists with a yen for nature are visiting rainforests and going on safaris in increasing numbers, and the money they spend on hotels and guide services often goes directly into the local economy. The business provides an incentive to maintain the wildlife areas in their pristine, natural state.

Tourism is what Leakey is banking on to save the elephant, Amboseli, and perhaps all of Kenya as well. Last year, nearly 700,000 tourists visited Kenya and spent roughly $400 million, surpassing coffee as Kenya's greatest source of hard currency.

The economics of tourism make sense. Since nearly all these tourists came at least in part to see the elphant, each of Kenya's 20,000 elephants is in effect responsible for bringing $20,000 into the country—ten times the amount of money its tusks are worth. What's more, notes Perez Olindo of the African Wildlife Foundation's Nairobi office, live elephants are a renewable resource. "When you shoot an elephant for its tusks, it is gone forever," says Olindo. "But a live elephant will bring in tourist money year after year."

In addition to boosting tourism, Leakey plans to take a more active role in managing the ecosystems in Kenya's parks. "Anyone who says you shouldn't manage a park is talking from another century," says Leakey. "With our population growth and development of agriculture, we're basically interfering anyway. So we might as well do it consciously."

At Amboseli, for example, Leakey is considering several steps to control the elephant population, such as using electric fences to block the elephants from some areas of the park, which would allow the trees to grow back and encourage the ele-

phants to migrate again to search for new food sources. There is even talk of reducing the number of elephants through selective killing, or culling, if the population gets too large.

Ultimately, the fate of Kenya's wildlife depends on the cooperation of people living near the parks. To that end, Leakey plans to make the parks a vital part of the local economy by guaranteeing that the people living nearby receive a greater share of the money generated by the parks to use for building schools and health clinics—and he promises to fix the water system at Amboseli.

Critics' Circle

To some conservationists, the idea of managing a park's ecosystem, culling herds of animals, and promoting tourism seems a cure worse than the disease. What is natural, they ask, about a wildlife park managed by and for the ultimate benefit of humans? Referring to a park in South Africa that has some 8,000 elephants supplied with a manmade system of water holes, one U.S. conservationist complains: "Kruger Park is nothing more than a big zoo."

For Leakey, there is no choice other than to actively manage wildlife. The alternative is to yield to those who see extinction as an inevitable part of evolution and conservation as futile. In that view, losing the elephants in Africa or even leveling thousands of acres of Amazon jungle is inconsequential when compared, for instance, with the cataclysmic disappearance of the dinosaurs 65 million years ago. At least 90 percent of the species that ever existed are gone, they note, so why worry about a few more?

But the real issue is not whether life on

WHAT YOU CAN DO

Don't buy animals or plants taken illegally from the wild. Before you buy a plant or pet, ask the store owner if it was captive bred or taken from the wild. If it was taken from the wild, ask the store owner whether he or she is sure the pet can be legally sold. Unless you ask questions, illegal trade will not decline.

World Wildlife Fund

earth will survive but whether the planet will remain hospitable to humans. For we have come to a point where the fragile ecological system that ensures mankind's survival is affected by our every action. As Leakey believes, man's concern for the survival of other species is in fact enlightened self-interest, whether it is saving a rainforest for its potential pharmaceuticals, a plant for its useful genes, or an elephant simply because it is 6 tons of regal beauty.

From *U.S News and World Report*, October 2, 1989. Copyright © 1989 *U.S. News and World Report*. Reprinted by permission.

EARTH CARE ACTION

Elephant Man

By William F. Allman

As a young man, Richard Leakey learned to fly airplanes so he could explore inaccessible areas of the African bush for promising fossil sites—and keep the rest of the world out of his hair. When he built makeshift airstrips in the bush near the sites of his excavations, the paleoanthropologist purposely made the runways short, knowing that pilots without his flying abilities wouldn't attempt to land there. "Unwanted visitors are a real nuisance," he says. "It is a very effective filter."

It is a philosophy that has guided Leakey all his life: Be your own pilot and build the runways short. His attitude has won him both admiration and irritation among his friends and colleagues, but the high standards he sets for himself buy him independence and filter out others whose shortcomings would only get in his way. The message of the short runway is clear: Everyone is welcome; all Leakey asks is that you be as committed as he is.

No Small Order

As the new director of Kenya's Department of Wildlife Services, the 44-year-old Leakey is usually at his Nairobi office by 6:00 in the morning, and he has no apologies for calling colleagues as late as midnight to discuss business. He rarely eats more than one meal a day, has little social life, and nowadays only occasionally sees his wife, Maeve, who is busy carrying on his excavation projects in the north. On his desk are four telephones, and most of the time at least one of them is ringing.

When Leakey inherited Kenya's wildlife department last spring, it was in shambles, with inadequate funding, widespread corruption, and decaying facilities. He has responded with a range of dramatic proposals, including firing nearly 2,000 park officials suspected of corruption, fencing off areas where wild animals and local farmers are in conflict, and turning Kenya's entire park system into a self-sustaining tourist business that channels some of the proceeds directly to the people living near the parks. Though some conservationists question his policies, no one doubts the influence Leakey has already had, if not yet in increasing the elephant population, at least in raising the current level of hope. "Without Leakey," says Iain Douglas-Hamilton, a Nairobi-based biologist with the World Wildlife Fund, "Kenya's elephants would be dead ducks."

The son of the late Louis Leakey and Mary Leakey, the celebrated paleontologists who discovered the fossils that placed the origins of man in Africa, Richard was determined at first not to continue in the family tradition. The Kenyan-born youth decided against a university education and instead started his own business as a safari guide. Eventually, he took to organizing the logistics of his father's excavations. Then, while accompanying his father to a meeting of the Committee for Research and Exploration of the National Geographic Society, he surprised the committee—and infuriated his father—by asking for funds to set up his own search for fossils near Kenya's Lake Turkana.

The committee gave him the money but cautioned him that, as a young man of 23 with no formal training in paleoanthropology, his reputation was on the line. Leakey's hunch that the area was rich in fossils paid off, however. Two decades of digging have yielded a wealth of ancient human remains crucial to piecing together our ancestors' early history. His discoveries, writes his mother Mary in her autobiography, *Disclosing the Past*, were "as spectacular and important as anything Louis or I had found." The finds gave him a degree of fame rarely enjoyed by scientists, landing him on the covers of news magazines and making him a best-selling author.

The success Leakey achieved in his old job has given him an edge in doing his new one. "I satisfied my ego a long time ago with fossils," he says. "I became someone who is known, listened to, and, in a sense, protected. And I think that is important to what I am doing now. I don't care what people think of me, and I don't care if I get sacked this afternoon. That's an enormous advantage." Though his efforts to revitalize Kenya's wildlife parks consume much of his time, Leakey is not giving up fossil hunting entirely. "Most corporate chiefs play golf," he says wryly. "If I skip the golf, I should be able to do a little anthropology."

Managing Nature

Leakey's career switch is in many ways fitting. Conservation has long been a Leakey family tradition. Richard is chairman of the

East African Wildlife Society, an organization started by his father, and his mother is a staunch conservationist.

While director of Kenya's National Museums, a position he held for two decades before taking on his new responsibilities, Leakey complained loudly that government corruption was abetting the poaching trade. It was a gutsy move in a country whose government is not particularly known for its tolerance of public dissent. As a result of his comments, he was accused by one government official of having a "cheeky white mentality" and was asked to resign the directorship of the museum, which he refused to do. Then last spring he found to his surprise that he had been put in charge of the very problem he was complaining about.

Despite the seeming dissimilarities of his old and new jobs, his present career takes advantage of what Leakey does best. A natural manager, Leakey has a passion as much for means as for ends. At his first dig as a young man, he was more interested in the logistics of the operation than in science. "I had enjoyed the challenge of getting the expedition to the camp and everything set up," he wrote in his autobiography, *One Life,* "but the fossil hunting and archaeology left me rather bored."

Nothing moves Leakey more than a seemingly undoable task, and, as difficult challenges go, he has taken on a daunting one. Faced with the threat of assassination (he is now accompanied by bodyguards), government corruption, and gangs of heavily armed poachers who have driven his native Kenya's elephant population to the brink of extinction, Leakey has responded to the changes in his life in characteristic fashion: He's getting the best

sleep he's had in years. "When I was running the museum, I was getting nine hours of sleep and felt exhausted. Now, I'm getting along fine on five."

Always adept at raising money for his archaeological projects, he is now embarking on an effort to raise some $150 million to revamp Kenya's wildlife parks, a remarkable sum in a country where annual per capita income is less than $400. But Leakey insists that, once his projects are up and running, the parks will pay for themselves as well as supply much-needed capital for Kenya's economic development.

Family Role

Leakey's love of Kenya drives his efforts as much as his concerns for wildlife. The Leakey family has long played a large role in Kenya's history; besides his father's work in starting the Centre for Prehistory and Palaeontology, part of Kenya's national museums, Richard's younger brother, Philip, is the only white member of Kenya's Parliament. "I'm going through a strong nationalistic phase in my life," says Leakey. "I've been asked to take on the number-one position in a sector of the economy that is crucial and in a mess, and that is exciting. The potential is enormous to turn this country around."

Because concern over the conservation of wildlife ultimately hinges on how we regard our species' place in nature, Leakey's previous career gives him an added perspective on humans' role in the history of life on earth. His excavations show that as far back as 2 million years ago, man's ancestors were turning nature to their advantage by chipping flakes of rock to use as cutting tools. "What separates us from other animals," says Leakey, "is the ability to think of tomorrow in terms of yesterday."

For Leakey, our species' ability to learn from the past to adapt ourselves to the future is a source of optimism. "Humans have always survived by managing their environment," he says. "There is no reason we can't do it now."

Leakey is careful to point out, however, that being a manager is not being a master. It is a lesson Leakey learned early in life. As a teenager, he once attempted to catch a leopard with a baited cage. When he went to inspect the trap that night, he found the cage empty but turned around to find a leopard blocking his escape. Leakey's only recourse was to enter the trap himself, causing the cage to snap shut, and wait until morning for help to arrive. It is a parable that might be applied to the challenges Leakey faces in his new job as well: He has set the trap, but the leopard is still at large.

From *U.S News and World Report*, October 2, 1989. Copyright © 1989 *U.S. News and World Report*. Reprinted by permission.

EARTH CARE ACTION

A Reason to Whoop

By Gary Turbak

On a cool morning in May, a single-engine Cessna cruises above the muskeg wilderness of Canada's Wood Buffalo National Park. From the passenger seat, biologist Ernie Kuyt keeps a close watch on the landscape scrolling by. Suddenly, up ahead, sunlight dances off something white. At Kuyt's gesture the pilot banks, surrenders altitude, and rushes toward the spot.

Standing out like jewels amid the marsh and willows and tamarack are two snow-white whooping cranes, their satin feathers and scarlet crowns glistening, their stately elegance belying the harshness of the habitat. One bird sits on a frail-looking nest of bulrushes. The other struts nearby, flapping its black-tipped wings in defiance.

At 90 mph the plane roars past, Kuyt's attention riveted on the cranes. "Get up, you," he orders, and as if obeying, the sitting bird stands to reveal its treasure. "Two eggs!" shouts Kuyt. The experienced pilot knows his cue, and the Cessna arches upward to strike a course for the next nest on Kuyt's list.

Soon, another pair of cranes unexpect-

At Wood Buffalo Park in Canada, Ernie Kuyt slides a whooping crane egg into a wool sock, an ideal egg tote. He visits every nest, making sure each contains one good egg; surplus eggs are hatched elsewhere. (Robert Semeniuk/First Light)

edly appears below—birds that Kuyt has not seen on previous flights. "These damn whoopers are everywhere," he growls with mock anger. "I'll soon be out of a job."

Ernie Kuyt (pronounced "kite") would love to work himself out of his job, which for the past 23 years has been to save the endangered whooping crane from extinction. Officially, he is the Canadian Wildlife Service biologist assigned to the crane project, but his involvement transcends the official. Over the years he has become the crane's key patron and protector, a sort of godfather to the great white birds.

Kuyt has discovered the secrets of

their nesting sanctuary, gently collected their eggs, soared with them, and defended them from threats ranging from bears to bombs. "The cranes have become his life and almost his religion." says Graham Cooch, former whooping crane coordinator for the Canadian Wildlife Service. With Kuyt's help, the whoopers have taken a big step back from the brink of extinction, in the process becoming one of the best-known wildlife success stories in North America.

Like the whooping crane, Kuyt is tall and lean—perhaps from years of eating his own cooking in northern outposts. Thick eyeglasses lend him an owlish, professorial appearance. Affable and avuncular, he has a quick grin and a wry sense of humor. His 60-year-old shoulders stoop a bit, and ailing knees cause him to walk with a bandy-legged shuffle. He speaks in measured cadence and with a whisper of an accent, a remnant of his Dutch upbringing. (He emigrated to Canada at age 18.) A consummate naturalist, he gets as excited about discovering a toad-hibernating area (which he did last spring) as other people would about winning the lottery. Mostly, though, Kuyt gets excited about "his" whoopers. "They're a perfect symbol of international conservation," he says, "a rare and beautiful creature with dual citizenship and an excellent future."

Dwindling Numbers

The whooping crane evolved during the epoch of the saber-toothed cat and the mastodon, on the wetland savanna that covered much of North America. Its haunting, buglelike call once rang across the continent, and probably neither the bird nor its cry has changed much over the intervening years.

Standing 5 feet tall (North America's tallest bird), with wings spanning 7 feet, whoopers are striking, majestic creatures. Though they appear fragile, they are actually formidable fighters that need fear only wolves, bears, and other large predators. Of the world's 15 crane species, the whooper is the only one other than the sandhill that lives in North America, and it is easily the rarest.

Long before the arrival of humans, natural forces had begun drying up the vast northern marshes, turning them to prairie. Over the centuries, crane numbers gradually dwindled. By 1850, there were probably fewer than 2,000 of the big birds in only a handful of U.S. states and Canadian provinces. As farmers drained the wetland habitat, the crane's nesting ground retreated northward and its numbers continued to shrink. Migratory whoopers have not nested in the United States since the last birds disappeared from Iowa in 1894.

In 1922, near Kiyiu Lake, Saskatchewan, game warden Fred Bradshaw stumbled upon a whooping crane nest, by then a rarity even in Canada. It contained a chick and two eggs. He calmly killed the young bird, probably to have it stuffed, and carried off the eggs as specimens. It would be 32 years before anyone saw another nest.

Though no one knew where they nested, the big birds tumbled almost magically out of the Texas sky every autumn to spend the winter in the warm coastal waters. Come spring, the whoopers flew north, but attempts to follow them to their nesting ground failed. As the years ticked by, the species' plight worsened; in 1941 only 16 whoopers showed up in Texas, and conservationists braced for its extinction.

Then, on a hot June afternoon in 1954, forester George Wilson was returning from a fire in the Northwest Territories when he peered down from the helicopter at an electrifying sight: two large white birds and a rust-colored chick. "My gosh, they're whoopers!" Wilson recalls saying to the pilot. The secret nesting area had been found—2,500 miles north of the Texas coast. Luckily, it lay inside Wood Buffalo National Park, a wild and inaccessible region larger than Denmark.

In 1937, the crane's wintering area

(now called Aransas National Wildlife Refuge) had also earned protection. But even with summer and winter sanctuaries, the crane population grew very slowly. In 1966, with the flock containing only 43 birds, Canadian and U.S. wildlife authorities decided to take an active role in the crane's life. Ernie Kuyt was already in the Northwest Territories studying wolves and caribou for a predator/prey study, so he drew the assignment. "I didn't really want the job," he recalls. "My wolf work wasn't finished, and the entire world was watching the crane project."

A Lasting Bond

It wasn't long before the reluctant biologist and the endangered crane struck up an enduring partnership. "I don't like to admit it," he says now, "but I've become emotionally attached to these magnificent birds." And although they don't know it, the cranes are dependent on him.

"It was extremely fortunate for whooping cranes that Ernie was assigned to the project," says Rod Drewien, a biologist with the University of Idaho who manages a whooper flock at Grays Lake National Wildlife Refuge in Idaho. "Few people could match the dedication and enthusiasm he has shown for these birds."

Because the whooping crane's muskeg breeding area is impenetrable from the ground, Kuyt accomplishes his field work via frequent flights from the tiny community of Fort Smith, where he maintains an office and spartan apartment. Two or three times a week during the spring he flies to the 400-square-mile nesting area to locate birds, note the distribution of pairs, and— most important—count their eggs. Later,

he meticulously records every tidbit of information and plots sightings on maps and aerial photos. It's all part of his mission to collect and relocate the eggs that otherwise might not survive.

Finding whooper nests in this maze of timber, bogs, and small lakes seemed impossible at first, but Kuyt has come to know these labyrinthine wilds the way most people know their own backyard. Even though all nesting whoopers stayed out of sight for decades, now he locates about 80 percent of the flock.

The focus of Kuyt's work has not changed since he was charged in 1966 with implementing a bold new plan. Because each whooper pair usually lays two eggs but rears only one chick (the other almost always dies, probably of starvation), authorities decided to remove some eggs for artificial incubation and hatching. What began as a controversial experiment has matured into a successful management technique that culminates each spring with Kuyt leading a raid on the whooper nests.

Last year's [1989] great egg snatch began just before noon on May 29, as Kuyt directed his helicopter across the wilderness to the first nest. At the chopper's approach, the adult cranes scurried from the nest. Upon touchdown, Kuyt and Jacques Saquet, warden at Wood Buffalo Park, slogged 100 yards through knee-deep water to the waiting eggs. Kuyt carried a pole for probing the treacherous marsh bottom, and Saquet toted a plastic bucket of warm water.

At the nest, Kuyt carefully floated one of the large buff-olive eggs in the bucket. The chick's movement inside rocked the egg slightly, a sign that it was a good one, and the men returned it to the nest. The second egg also contained a live chick, but Kuyt slipped this one into a wool sock for transport back to the helicopter. At the chopper, he labeled the egg with a soft lead pencil and placed it in a padded suitcase kept warm with hot-water bottles. Minutes later, the helicopter was on its way once again, and the adult cranes had returned to their nest.

Each year, the collecting continues until every nest has been visited and all surplus eggs removed. If both eggs in a nest are bad, Kuyt replaces them with a good one from the suitcase. (Last year, Kuyt performed his egg sleight-of-hand at 28 nests.) By day's end, every wild whooping crane nest in the world holds one egg with a live chick inside. Egg-swapping has increased hatching success in the Wood Buffalo flock by nearly 20 percent.

Surplus eggs are hatched artificially at Patuxent Wildlife Research Center in Maryland or slipped into sandhill crane nests at Grays Lake in Idaho. By raising the adoptees, the sandhills have helped to establish a second wild whooper flock. The Grays Lake area is home to about 20 whooping cranes, but because none so far has produced any young, U.S. and Canadian wildlife officials have considered abandoning the program. About 50 captive whoopers lived until recently at Patuxent; last summer, scientists decided to move half of that flock to the International Crane Foundation at Baraboo, Wisconsin.

Handle with Care

Kuyt conducts the egg collection with military precision, but though he has never lost an egg, there have been glitches. The first egg pickup took place in 1967 amid fears by some conservationists that the experiment might somehow harm the whoopers. "We made it very clear to Ernie that *nothing* must go wrong," recalls Graham Cooch. One important preparation was the construction of a special plastic foam case for carrying eggs to the helicopter. As the chopper neared the first nest, however, panic jerked Kuyt upright in his seat. He had forgotten to bring the case, and going back for it would throw the whole operation off schedule. An alternative had to be found—and quickly. Then he remembered the spare wool socks in his pocket. The socks worked so well for carrying eggs that the high-tech carrying case has never been used.

During the 1980 egg collection, Kuyt encountered a black bear sloshing its way boldly toward a nest. Rather than retreat and surrender the nest, the biologist picked up a stick and, with a well-placed toss, beaned the bruin, driving it away. "It's no secret that I feel protective and selfish about whoopers," Kuyt says. "But they need all the help they can get."

The birds face untold perils. Power lines and storms kill some cranes, and last January a Texan shot one of them. In all, 19 whooping cranes were lost in the 1988–1989 migration period. Whooper watchers worry most about the Gulf Intracoastal Waterway, a shipping canal running through the Aransas refuge. Since its completion in 1940, canal erosion has destroyed 1,150 acres of crane marsh and continues to swallow up 2 additional acres a year. "What's really scary," adds refuge biologist Tom Stehn, "are the barges that transport oil and toxic chemicals just 100 yards from the cranes."

Despite the dangers, whooping cranes

have made a remarkable comeback—though not so great as to eliminate Kuyt's job just yet. Two years ago, 148 whoopers flew south from Wood Buffalo Park, an 825 percent increase over the 1941 low. Most of the gain has come this decade, with the flock doubling since 1981. Recently about 20 new chicks have joined the population each year, and when these birds start reproducing at age four or five (their life span is 20 to 25 years), the flock could increase significantly. "I hesitate to say so," says Kuyt, "but the population may be on the verge of mushrooming.'

Sharing Flight

For him there has been no greater joy than this whooper boom, except perhaps the time in 1981 when he tagged along with the birds on their southward migration. That summer, he and Rod Drewien attached tiny radio transmitters to three chicks. When the whoopers journeyed to Texas, Kuyt—the man who would be crane—flew with them.

Whoopers travel in families or small groups and make many stops in their month-long journey to Texas. The return trip in the spring, with fewer rests, may take only 11 days. Cranes flap their wings very little during migration. Instead, they ride warm air currents upward, then glide down in the direction they want to go—traveling up to 600 miles a day.

With receivers mounted on a small plane, Kuyt monitored—and mimicked—the birds' every movement. When they flew, so did he. When they stopped for the day, he touched down at the nearest landing strip. For 2,500 miles, Kuyt and the cranes were like one. Never before had anyone tracked migrating birds so far, or so closely. With the help of skilled pilots, Kuyt was often able to keep the cranes in sight as they pressed southward. "To soar with the whoopers, to share their airspace, to be like a crane on migration was a fabulous experience," he says. "After a while, I began to think like a crane."

Once, he had to think *for* them. Late one afternoon over Oklahoma, with three whoopers in sight ahead of the plane, Kuyt was alarmed to see explosions not far away—bombing and gunnery practice at Fort Sill Military Reservation. With the cranes headed directly into the line of fire, he frantically radioed the base commander, and to his surprise, the shelling stopped almost immediately. The plane, however, had to fly around the restricted military zone. Like a worried, pacing parent, Kuyt flew back and forth along the reservation's southern boundary until the cranes emerged. When the birds were safely out of range, the shelling resumed.

Kuyt looks forward each year to the northward migration—the cranes' and his. The initial reconnaissance flight from Fort Smith in the spring is a special event, a trip he makes alone. "After all these years, it's still exciting to spot the first returning whooping cranes, to know they made it back safely," he says. "It's like greeting old friends."

From *International Wildlife*, January 2, 1990. Reprinted by permission.

Bringing the Birds Back to Assisi

By Cynthia Hanson

Every September in Florence, Italy, the Fiesta Degli Uccelli (Fair of the Birds) ushers in the new hunting season. The Viale Machiavelli is blocked off, and hunters display guns, ammunition, and wild songbirds in cages (used to attract free-flying birds).

But this year the only unusual sight on the street were some black-bordered notices lamenting the "death" of the fair.

Hunting in general, and the hunting of songbirds in particular, is under fire in Italy (which is located on a major songbird migration path).

Throughout the country, conservationists rally for revised hunting laws. Petitions circulate, politicians speak out. A Rome-based newspaper poll in December 1986 found that 80 percent of the public wanted some reduction or the total abolition of hunting in Italy.

Watching from the sidelines is American Bert Schwarzschild, who began a songbird campaign in Italy in 1983. Schwarzschild played a significant role in prompting the Italian public to rethink a tradition he calls "barbaric" and "unsporting."

Songbirds are a delicacy in Italy, and are roasted, baked in bread, and prepared in other ways. Schwarzschild's concerns led to a national campaign in Italy, and the establishment of the Assisi Bird Council/ Europe, which he cofounded with Maria Luisa Cohen in 1984.

Rousing thought is practically second nature for Schwarzschild, an electrical engineer-turned-environmentalist. "I've always had an empathy for the plights of other people and the plight of other living beings," says the soft-spoken, Berkeley, California, activist. He was the national director of the Whale Center in Oakland, California, executive director of the American Youth Hostels, and chairman of the U.S. education committee for the International Union for the Conservation of Nature and Natural Resources. He also founded an American branch of the Assisi Bird Council.

A Cruel Fate

Schwarzschild found the impetus for his Assisi Bird Campaign while vacationing in Italy in 1983. He was intrigued with the idea of hiking up Mt. Subasio, the place where Saint Francis of Assisi befriended birds and animals in the thirteenth century.

"As I started walking up, I heard shotgun blasts in the valley below," Schwarzschild recalls. "Then I started to notice hundreds of shotgun casings on the trail, and feathers in the bushes. And I realized I didn't hear a single songbird . . . The songbirds of Saint Francis were no more and had been decimated by hunters."

According to the International Council for Bird Preservation in Oxford, England, hundreds of millions of songbirds (perhaps 15 percent of those migrating) are shot, trapped, netted, or caught with lime-sticks (branches that are coated with a sticky substance) as they travel the Mediterranean flyway. Some of the hunting is legal, some is not. Italy has the second-largest concentration of hunters in the world, after Malta.

"There's a war going on between hunters and protectionists," says Dr. Yves Lecocq, secretary general of the Federation of Hunters' Associations for the European Economic Community (FACE) in Brussels, Belgium. "It's not by banning that the situation will improve," he says. He would rather talk to the hunting groups and get them to "intensify their education for members."

FACE, which lobbies the European Economic Community on behalf of hunters, condemned Italy in 1987 for hunting 11 species of birds and mammals protected in other European countries. Italian hunting laws are permissive and poorly enforced, says Dr. Lecocq.

Lecocq is quick to defend Italian hunters, however, and says songbird hunting is "often an emotional issue." "We think that [hunting] should only be discussed on a scientific basis, and as long as it can be viewed as wise use, nothing is wrong with it.

"According to research, more birds are killed in the United Kingdom and Germany by domestic cats than birds in Italy," he says. "But it doesn't excuse what is

sometimes happening in Italy" with illegal hunting, he says.

Besides launching a campaign to save songbirds, in 1983 Schwarzschild published his experience in *Audubon* magazine. In it, he recommended a moratorium on hunting songbirds on Mt. Subasio. A torrent of mail opposing the hunt poured into the town of Assisi—so much that the mayor had to have his mail delivered in a wheelbarrow, says Schwarzschild.

The issue of shooting songbirds spread to Italian newspapers and magazines. A "bird day" was celebrated in Assisi with musicians and a symbolic releasing of doves. And the newly elected minister of the environment, Alfredo Biondi, attended the celebration and declared that the birds should be brought back to Assisi. The issue became a national controversy that continues today.

Schwarzschild appealed to Italian pride: "How can you, a civilized country with your traditions and your monuments of compassion—because the Italians by and large are very compassionate people—how can you allow this to happen?" he asked them.

"I can safely tell you that we were a catalyst in bringing together the conserva-

WHAT YOU CAN DO

Save some space for wildlife. The fight to conserve wildlife goes on all over the world. This includes your own backyard. There's a lot of wildlife there, from plants to insects to birds. Protect it!

World Wildlife Fund

tion organizations in Italy who had been discouraged because the hunting lobby in Italy is so powerful," says Schwarzschild. "It was our effort that got the media worked up."

A moratorium has been placed on songbird hunting on Mt. Subasio, and the Italian government recently approved the

boundaries of a park on the mountain, though no funds for the park have been set aside.

EARTH CARE ACTION

Citizens Do Their Dolphin Duty

By Kristine F. Anderson

"Dolphins at 12 o'clock," calls skipper Ed Flinn as he guides his boat along the mouth of the Savannah River near Savannah, Georgia. "There's three or four just 20 feet ahead of us."

The crew scrambles to the bow as the small pod, or group, of dolphins swims along the water's surface. "Listen," says Gwen Flinn, the skipper's wife, as she quickly focuses her camera, "You can hear them chuffing and squeaking. They're puffing through their blowholes."

"Dolphins at 3 o'clock," yells Dawn Averitt, a student at Georgia State University. The crew members quickly turn to watch another group of dolphins perform their acrobatic feats, and Averitt begins to record the latitude and longitude, approxi-

mate water temperature, number of adults and calves, and any unusual behavior.

Pausing to watch the playful mammals' antics, she says, "See that one with the white face—he's looking right at us and smiling. He's just as curious about us as we are about him."

Pointing to the second pod sighted, Debbie Snider, a hairdresser from Savannah, excitedly asks, "Is that a mother and her calf? And look, there's one with a clipped dorsal fin—he must have gotten caught in a net or something."

A Ten-Year Study

The Flinns and their crew are part of an Atlanta-based citizens group working to

Dolphins, well known for their acrobatic abilities, jump and spin off the Florida coast. In nearby Georgia, volunteers help scientists count and catalog these highly intelligent animals. (Greenpeace/Visser)

help scientists and government officials count and catalog Atlantic bottlenose dolphins. They participated recently in the second survey effort, which extended over a 390-square-mile area along the Georgia coast. Like the 200 other volunteers involved in the weekend survey, they scanned the area, including nearby tidal rivers, bays, and creeks for signs of approaching dolphins.

The Dolphin Project, a nonprofit organization made up of close to 300 citizens, has embarked on a ten-year research study to gauge the effects of the 1987–1988 epidemic that killed off an estimated 50 percent of the dolphin population along the Atlantic coast.

The project's advisers believe that they can learn more about the overall marine environment by observing the recovery of these intelligent creatures.

Charles W. Potter of the Smithsonian's Marine Mammal Program and the chairman of the board of scientific advisers for the project, says, "We found massive amounts of pollutants in their system when we examined dead dolphins. Since dolphins are at the top of the food chain, their food source is the same as ours. We ingest many of the same pollutants and toxins they ingest every time we eat sea trout, mullet, or croaker."

While dolphins are not threatened with extinction, surveyors hope to monitor the expected recovery of the dolphins from the "die-off," and if necessary call for government intervention and stricter protection of dolphins and coastal areas.

Volunteers, who will participate in four surveys each year, are working closely with Potter and several other scientists and advisers. Since the project is still in its early stages and so little is known about dolphins' behavior in their natural habitat, volunteers are trying to collect as much data as possible.

Beau Cutts, the group's president and a journalism professor at Georgia State University, says, "One of the payoffs [of the project] is that we can observe seasonal changes and migration patterns to get some meaningful data."

Potter adds, "I think the concept of a volunteer group composed of enthusiasts working in conjunction with professional dolphin researchers is an excellent way to provide the general public with a feeling for the animals and their role in the environment."

Volunteers range in age from the late teens to retirement age and come from all over the Southeast. Though most work full time in a variety of jobs ranging from computer programmer to psychotherapist, they share a particular affection or concern for dolphins and a desire to get involved in a hands-on effort.

Pat Cote, the group's vice president and a senior data system consultant in Atlanta, says, "I've done a lot of volunteer work in the past ten years, but with this project I have a real chance to accomplish something and do more than just write a check."

While the project is less than a year old, it's the most comprehensive long-term study of dolphins anywhere in the world because of the large geographical area surveyed and the extended time period. Cutts

is confident that volunteers can cover the whole Georgia coast in the next few years. He knows of two smaller groups doing surveys in North Carolina and says that he's had inquiries about starting a similar group in Florida. He hopes a volunteer network can be organized within the next year or so.

A Working Model

According to Gerald Scott of the Miami laboratory of the National Marine and Fisheries Service, if the dolphin project is successful in Georgia, it will become a model for projects in other states.

During the first survey last July, 30 teams of volunteers sighted 1,374 dolphins. In mid-October, volunteers sighted only 559 dolphins, even though record-keeping procedures had been refined. Cutts says, "The lower number could have multiple causes, including the falling tides [on the second-survey day in October], migration patterns, and the change of seasons."

Jill McKenzie, a criminal justice major at Georgia State University who participated in both surveys, was on one of the teams that didn't sight any dolphins. She says, "We were disappointed, but we can learn just as much from what we don't see as what we do see. Hopefully, as the survey continues, we'll know why we didn't see any."

While the data collected from the first two surveys is still being complied, Cutts says, "The results of the survey will have significance beyond Georgia, particularly in terms of long-term migration and reproduction and any signs of disease."

Volunteers pay $25 annual dues, a $72 survey registration fee, and their own travel expenses to Savannah for the weekend surveys. They also help pay for the fuel for the small boats.

Cutts estimates that each survey costs between $12,000 and $14,000, but he quickly notes that if the government attempted a similar effort it would likely cost close to $130,000. The federal government did provide some technical advice in helping the group collect data and also provided nautical charts. Cutts hopes to get some federal and state funding as well as grant money in the next year.

Along with surveying and photographing the dolphins for the catalog, the volunteers are also educating others about the man-made and natural problems threatening these benign creatures.

Along with other environmental groups, the Dolphin Project has endorsed a resolution introduced by Rep. Barbara Boxer of California. The resolution, the Dolphin Protection Consumer Information Act, deals with the slaughter of dolphins by tuna fishermen and requires that tuna be labeled according to the technique used to capture them.

The Dolphin Project is also asking consumers not to buy any yellow-fin canned tuna, but to instead buy white albacore tuna. (The albacore tuna is caught with a hook and line rather than in purse nets.)

Cutts sums up the project: "Ordinary people can make a difference—we don't have to wait for the government or big business to act."

EARTH CARE ACTION

A Pelican's Best Friend

By Wayne Mayhall II

Ten years ago, suffering from worsening heart problems, Dale Shields reached a low point in his life. "I felt like I had nothing to show for myself," he recalls, despite having six grown children, a successful 25-year career as a salesman in the Midwest and a new home and wholesale produce business in Sarasota, Florida. Shields even stopped enjoying one of his favorite hobbies, fishing. Then one hot summer day, a single injured bird changed his life forever.

"I had planted myself on a dock and was casting out my fishing line," he says, "when I saw a brown pelican lying on the rocks." Moving closer, Shields thought the bird was dead until he made eye contact with the creature. "He was almost gone but not over the edge. I began calling people from Tampa to Key West but no one knew what to do. So I took the bird into my own care and somehow it survived."

Shields has been helping birds—

Dale Shields says his fight to save pelicans is what keeps him alive. (Bill Wisser)

brown pelicans in particular—survive ever since. During the past decade, he has been the motivating force behind the rescue and rehabilitation of more than 16,000 birds along the Florida Gulf Coast. That motivation, says the 63-year-old survivor of seven coronary bypasses, is the reason he himself is alive today.

Shields' nonprofit Protect Our Pelicans Society, founded in 1985, now has more than 4,000 members, including nearly 200 Gulf Coast volunteers who are available around the clock. Shields, meanwhile, is licensed in rehabilitation education by both state and federal authorities and spends much of his time these days lecturing throughout the state to civic and school groups. His efforts have led one community, the town of Longboat Key, to hold an annual "Pelican Man Week." And he was honored recently as Sarasota's Outstanding Senior Citizen.

"There may be few things we politicians can agree on, but Dale Shields and his efforts to help the pelican is one of them," says Jim Greenwald, chairman of the Sarasota County Commission.

Wildlife experts frequently question the value of bird rehabilitation efforts, arguing that such programs have only minimal impact on animal population and do nothing to help solve the big problem of habitat destruction. No one disputes the notion, however, that the work of people like Shields does draw important attention to the plight of wildlife. And in the case of a species such as the brown pelican, which is considered endangered in many parts of its range, rehabilitation may help maintain the birds' overall numbers.

The normal life expectancy of a pelican is about 30 years. But the average life of one of the birds in Sarasota Bay, says Shields, is 7 years. One of the main reasons is that nine out of ten pelicans in the bay become hooked by fishing lines.

"Uninformed fishermen can make a deadly mistake," he says. "Instead of reeling in the entangled bird and removing the barb and line, they cut the line and let the pelican fly free." And all to often, once the bird returns to its roost, the line becomes caught and the creature hangs itself. The task of Shields and his army of volunteers is to get there before that happens. Last year, they did so successfully some 3,000 times.

From *National Wildlife*, October/November 1990. Reprinted by permission.

April 22, 1990—A huge crowd fills the Great Lawn of Central Park for a concert and rally marking Earth Day. (Chester Higgins/*New York Times*)

IN THE NAME OF EARTH DAY

By Denis Hayes

In 1970, we had the first Earth Day and somewhere between 20 million and 25 million people took to the streets. We had a wonderful window of opportunity through which we drove a lot of vehicles. When I look back upon these last 20 years, it is hard to think of a really important battle that the environmental forces did not win. And yet, those of us who set out to change the world are poised on the threshold of utter failure. Measured on virtually any scale, the world is in worse shape than it was two decades ago.

How could we have fought so hard, and won so many battles, only to find ourselves now on the verge of losing the war? The answers are complex. But if we can understand the mistakes that led to our current dilemma, we may yet be able to redeem our youthful promises to the next generation.

To start with, we were occasionally blindsided. Problems snuck up on us before anyone recognized the threat they posed. Even where there was agreement among the experts, the consensus was often later found to have been wrong.

In 1970, for example, nobody guessed that chlorofluorocarbons (CFCs) were going to be destroying the ozone layer. In fact, if you'd asked the average industrial chemist to give us a true success story, I think CFCs would have been on most people's short list—nontoxic, noncorrosive, nonexplosive, nonflammable. Ozone threats are not a unique example of our ignorance. Until recently, we employed asbestos routinely throughout our built environment, never dreaming the havoc it could cause to human health.

Some part of it was simple ignorance. But I think that is a very small part of it.

A somewhat more important thing is that we have repeatedly fought each battle on its own without a strategic overview. Despite all the environmental literature describing how everything is connected to everything else, we have ignored this elementary truth. In case after case, we have solved one problem by making some other problems worse or by creating some brand new problem.

We face a serious possibility of making the same error again. For example, some

197

are advocating biodegradable plastics as the answer to the plastic litter problem. Discarded six-pack holders can strangle birds and other species; plastic diapers are clogging our landfills. But the problems posed by biodegradable plastics are themselves serious. For example, when biodegradable plastic is mixed with other plastics, it renders the latter virtually impossible to recycle.

On a third front, we failed to set priorities. Among countries, we are what boxers call a counter-puncher. What we do best is respond. What America does not do well is anticipate and avoid problems. Unfortunately, many environmental phenomena involve thresholds that, when passed, cause damage that is essentially irreversible.

We face numerous such thresholds in the years ahead. Some, such as rainforest destruction, are already causing irreversible harm. Every area cleared is lost forever. Global warming could result in rising oceans covering huge tracts of land, including the rice-producing river deltas of East Asia. These are not problems we can respond to. They are problems we must avoid.

Fourth, we have ignored, as an environmental movement, some very important tools.

Idealism versus Realism

There was a huge contest back in the early 1970s between the people who viewed themselves as "lifestyle environmentalists" and those who were "political environmentalists."

In the passion of the times, they loathed one another. The political folks dismissed the lifestyle folks as naive idealists who wanted to retreat to a commune in Vermont and lead lives of irrelevance. The lifestyle people thought of the political folks as ones who wanted to go into smoke-filled rooms and engage in a process, the ultimate goal of which was to reach a satisfactory compromise, and to compromise issues in which compromise was simply unacceptable.

What happened was that the political folks won. If you give money to an environmental group today, 99 cents on the dollar goes to try to influence government. It goes for lobbying, it goes for election, it goes for litigation, it goes to try to affect rule-making processes.

But most environmentalists are willing—even eager—to do more than send money and write letters. We need to appeal to them as consumers, as workers, as investors, and as parents.

All the most successful movements have succeeded in part because they ask people to improve their behavior. The civil rights movement and the women's movement, for example, ask their supporters for heroic changes in their personal life. U.S. environmentalists, on the other hand, have often tried to convince the public that we could have our cake and eat it, too. People were encouraged to believe that, if we could affect the necessary changes in government and industry, we wouldn't have to change habits at all.

The answer to air pollution, for example, was claimed to be catalytic converters on tailpipes and scrubbers on smokestacks. We have pursued this strategy, at enormous cost, for 20 years. Yet the sky today in Los Angeles resembles split-pea soup. It was necessary, but not sufficient, to scrub pollutants out of exhaust. We must also begin using cleaner fuels and more efficient engines. We also should encourage widespread use of public transportation. We should also promote bicycle riding wherever possible. We should create incentives for people to live closer to their workplaces to reduce urban traffic.

Similarly, we should encourage environmental shoppers to be mindful of their values when they go shopping. The so-called green consumer has become enormously important in Europe.

We have not done much with green investment. We should have pioneered, perhaps, the path that the South African campaigns have followed of trying to influence people at board meetings, through voting of stock. A group that I co-chair has produced what we are calling the Valdez Principles for environmentally responsible corporate behavior.

I can go on and on with these tools. The point is that the environmental movement has not really used any of them, except the political tools, and if we are really in the kind of crisis that I think we are in, we cannot afford to ignore any of them.

The battles that we have won over the last 20 years have often been brutal. They have been hard. But, in retrospect, they have been the easy battles.

The last point that I will make is perhaps the most important.

The battles that we have won over the last 20 years have often been brutal. They

have been hard. But in retrospect, they have been the easy battles. There are the Clean Air Act, the Clean Water Act, Superfund. These are things that have price tags on them, to be sure. But they were things where a fraction of the population, enjoying at least shallow but very broad-based public support, was able to be successful in affecting the legislative process.

The battles that we are facing ahead cannot be won by an environmental movement that has 10 million or 12 million members and vague public support. The threat of global warming, for example, requires that we move swiftly away from fossil fuels and on to renewable fuels. It demands a transition from oil and gas to solar/hydrogen, universal use of passive solar architecture, and perhaps a trillion-dollar investment in energy efficiency. Such a transition will necessarily entail winners and losers. Conventional energy producers will be among the losers. These energy producers are some of the richest and most powerful institutions in the country and they will fight like hell to avoid being phased out of existence. Such transitions will be painful; they will be expensive; and they will require enormously broad-based public support.

And yet, the environmental movement has not diversified. Poor people and people of color are downwind from most toxic incinerators. They are down-gradient from most hazardous waste dumps. They are in the fields when the pesticides are sprayed. They work in factory jobs having the highest exposure to dangerous substances. Yet poor people and people of color are not well represented in our ranks.

Instead of having conferences where environmentalists talk to other environmentalists and the Sierra Club and the Audubon Society negotiate difficult sorts of compromises to get on the same wavelength, we need to reach out aggressively to sectors of society who have good reason to be environmentalists but who have not traditionally been part of this movement. We need to go to bat for them on the goals that they have, and show them why our goals should be important to them.

I guess that is the final point to make about Earth Day. When we put Earth Day 1990 together, we got the leader of every major environmental group in the country on our board. But what is exciting about the board is that we have leading political figures. We have leading scientists. We have civil rights leaders. We have leaders of women's organizations. We have labor union leaders. We have chairmen of major corporations.

We are going out to every sector of society, and I hope in the process of doing that are putting together an event that will convert America into a nation of environmentalists, almost to the point where the label becomes meaningless.

From *Environmental Action*, March/April 1990. Reprinted by permission.

EARTH DAY 1990

Vital Statistics

By Chris Wille

How does the health of the nation's environment and wildlife today compare to the time of the first Earth Day? The most obvious answer, of course, is that such comparisons are not easily made. For one thing, 20 years ago we did not know how to correctly measure all of the problems we faced; the same can be said for many of the dilemmas we face today. And all too often, the indices we develop for measuring different environmental problems become obsolete once we discover how extensive and complicated these problems really are.

Statistics do tell a story, however.

What the following numbers reveal is that considerable change has taken place in several areas in recent years—some good, some not so good. This eclectic collection of numbers is meant to be neither an indictment nor an accolade. Rather, it is merely a point of reference—figures for us to contemplate as we begin the 1990s, a time we surely will come to know as the "Decade of the Environment."

➤ World human population in billions in 1970: 3.72. Projected for 1990: 5.32.

➤ Number of national wildlife refuges in the United States in 1970: 331. In 1989: 452.

➤ Estimated global pesticide sales in 1975: $5 billion. Projected for 1990: $50 billion.

➤ Number of whooping cranes in existence in 1970: 71. In 1989: 217.

➤ Number of California condors in 1986: 60. In 1989: 30.

➤ Number of dusky seaside sparrows in 1970: about 1,000. In 1989: 0.

➤ Number of beverage cans used in America in 1963: 11.5 billion (mostly steel). In 1985: 70 billion (mostly aluminum).

➤ Of every federal dollar spent, the amount directed toward natural resources and the environment in 1976: 1.5 cents. In 1987: 3 cents. In 1989: 1.5 cents.

➤ World military expenditures (in 1984 dollars) in 1970: $450 billion. Projected for 1990: $750 billion.

➤ Estimated number of U.S. wetland acres lost in 1970: 500,000. Projected for 1990: 300,000.

➤ Miles of designated U.S. Wild and Scenic rivers in 1970: 868. In 1989: 9,278.

➤ Billions of board-feet of timber harvested from U.S. Forest Service lands in 1970: 11.5. In 1988: 12.6.

➤ Median age of U.S. population in 1970: 27.9. Projected for 1990: 33.

➤ U.S. population served by municipal wastewater systems providing secondary treatment or better in 1960: fewer than 10 million. In 1984: more than 125 million.

➤ Millions of tons of solid waste generated in the United States in 1970: 100. In 1986: 158.

➤ Millions of dollars appropriated by Congress from the Land and Water Conservation Fund to buy parkland and wildlife habitat in 1970: 48. In 1986: 45.9. In 1989: 207.

➤ Parts per million of DDT in human adipose tissue in the United States in 1970: 8. In 1983: 2.

➤ North American population of breeding mallards in 1970: 10,379,000. In 1989: 6,119,000.

➤ Number of pronghorns in North America in 1964: 386,000. In 1983: 1,051,500.

➤ Total U.S. energy consumption (excluding wood) in quadrillion BTUs in 1970: 70. In 1985: 73.

➤ Number of operating civilian nuclear power plants in the United States in 1970: 15. In 1989: 100.

➤ Total world carbon emissions in millions of metric tons from burning fossil fuels in 1970: 3,934. In 1986: 5,225.

➤ Number of whales killed worldwide in 1970: 42,105. In 1989: 300 (estimated).

➤ Number of U.S. homes using passive or active solar energy in 1970: 35,000. In 1987: 1,700,000.

➤ Millions of tons of sulfur dioxides emitted into America's air in 1970: 27. In 1985: 21.

➤ Thousands of metric tons of lead polluting America's air in 1970: 204. In 1985: 21.

➤ Number of states with working bottle bills in 1971: 1. In 1989: 9.

➤ Number of catalogued pieces of artificial space debris (softball-size or larger) counted by Space Command in 1970: about 2,000. In 1987: 6,985.

➤ Millions of acres of U.S. agricultural land transformed into urban areas between 1970 and 1980: 13.

➤ According to analyses of government reports, the amount of federal rangeland that was overgrazed and in "poor to fair"

condition in 1977: 70 percent. In 1989: 70 percent.

➤ Estimated number of African elephants in 1970: 4.5 million. In 1989: 500,000 to 650,000.

From *National Wildlife*, February/March 1990. Reprinted by permission.

 EARTH DAY 1990

Millions Join Battle for a Beloved Planet

By Robert D. McFadden

Two decades after the birth of the modern environmental movement, millions of people around the nation and the world renewed the call to arms for an endangered planet yesterday [April 22, 1990] with an exuberant and bittersweet celebration of Earth Day 1990.

It was not quite The Day the Earth Stood Still, as an avalanche of publicity in recent weeks had prophesied, but by day's end organizers of the sequel to Earth Day 1970 said 200 million people in 140 nations had taken part in the largest grass-roots demonstration in history.

And a vast patchwork quilt of marches, rallies, concerts, festivals, street

fairs, and other events in 3,600 communities in the United States, Europe, Asia, Africa, and North and South America seemed to bear them out.

"It's like Earth's birthday," said Gordon Vestnys, a 25-year-old roofer, strolling at a fair overlooking San Francisco Bay and the Golden Gate Bridge. "It's a gathering of people becoming more aware—and hopefully continuing after today with more consciousness and less Styrofoam."

From Italy to Nepal

Diverse, contradictory, extravagant: The bewildering global panoply of happenings

In a dawn ceremony outside Bozeman, Montana, Bill Tall Bull faces east to greet the rising sun as he performs a traditional Cheyenne Indian ceremony to cleanse and renew the spirit of a Cheyenne woman. The spring ritual was held in conjunction with Earth Day ceremonies. (Doug Longman/*New York Times*)

included a trash-picking trek up Mount Everest in Nepal, a 500-mile human chain across France, a gathering of space travelers at the United Nations, an "eco-fair" on a landfill island in Japan and a roadway "lie-down" by 5,000 people to protest car fumes in Italy.

In New York, Los Angeles, Chicago, San Francisco, Atlanta, and cities and towns across this country there were tree plantings, litter sweeps, recycling exhibits, workshops on wildlife and endangered species, fairs, and sunrise-to-sunset clamor for environmental morality that often mimicked the hoopla of a political campaign.

President Bush went fishing in the Florida Keys and used a telephone hookup to address thousands at an Earth Day rally in the Columbia River Gorge in Washington State. Members of Congress, governors, mayors, and hosts of other officials got on the bandwagon.

But it was not a day for politicians or corporations, although many of them embraced the common good with solemn pronouncements upon polluted air and oceans, fouled rivers and beaches, the perils of acid rain and tropical deforestation, toxic wastes and mountains of garbage.

"Start Saving the Earth"

Rather, it was a day for ordinary people and the meanings they chose to give it, a day to turn away from the memories of cold war and a hard focus on economic growth, to gather in vast throngs or quiet settings and express something about themselves and their world.

"It's really necessary to start saving the earth," said Kathy Bernstein, 60, at a festival in Chicago's Lincoln Park. "I've got grandchildren, and it's necessary to imbue in them that you have to save things, not just be users."

At a fair in the Playa Del Rey district of Los Angeles, Leslea Meyerhoff, 22, explained: "I'm here because I think individuals can make a difference if they are aware of recycling and conservation. One way to let people know of the environmental crisis is through activities like this that promote public awareness."

From some, like Philip C. Martinez, a drum-maker from the Taos Pueblo Indian Reservation in New Mexico, the urgency was greater. "We're here to save our land," he said at a rally in Washington, D.C. "They're trying to put both an airport and a nuclear waste dump near our reservation. Ten years ago, if we had said anything about it, nobody would have listened. Today, people care."

And in London, Andrew Lees of the British branch of Friends of the Earth said, "If Earth Day does nothing else, it will give the clear message to politicians that millions of people are aware of the problems facing the earth and what needs to be done about them."

The Weather Seemed to Approve

It was a balmy spring day in most of the United States. The sun arched over the continent in a sky that was pristine blue in the east and resembled a summer sea flecked with whitecaps in much of the rest of the country. Not everyone, however, turned out because of the day's gentle beauty or a commitment to the environment.

"I'm here to see Bob Weir—I'll admit it," said Matt Herren, 18, at a blustery, rainy San Francisco Earth Day concert featuring the Grateful Dead star. And across

the nation in Times Square, scene of an observance early yesterday, Jeffrey Aym, 21, of Morristown, New Jersey, acknowledged: "We're here because we missed Woodstock," the peace movement's 1969 rock-and-drugs festival.

But most people said they had come out because they believed their presence would make a difference, and the turnouts in many communities appeared to underscore a new political commitment to the environment by Americans, one coinciding with the fading era of confrontation with Communism.

Evocations of the 1960s

Some said the environmental movement has more support than any since the anti-war and civil rights protests of the 1960s.

And indeed, scenes reminiscent of the 1960s were abundant across the country. In New York, a huge throng—the police said 750,000, but it was like counting jelly beans in a jar—heard a concert in Central Park. In Washington, 125,000 people crowded onto the sun-drenched Capitol Mall: students in tie-dyed T-shirts and young professionals, older people who remembered the marches of a generation ago. Music and Frisbees floated up, and kites rode the breeze.

Many across the land felt no need to attend spectacular events. They stayed home or met in small groups to plant flowers or saplings, clean up roadsides and vacant lots, or just talk about their neighborhoods or why the loss of the ozone layer or the destruction of the Brazilian rainforests were important.

It was possible for some, away from

the crowds and the noise, to observe scenes of decorous beauty and be aware of the hidden pollution: sunlight shimmering like gold coins on the Connecticut lake choked with algae; vegetables banked at a stand in New York, an artist's palette of oranges, grapes, and bright red tomatoes coated with pesticides; and the Hudson River, tainted with polychlorinated biphenyls (PCBs), flowing in stately procession like a doomed blue regiment moving South.

Sending a Message

Such awareness was the essential point of the day, its organizers said. "It is incumbent upon each of us to make environmentally sound choices throughout our daily lives," said Denis Hayes, the 45-year-old chairman of Earth Day 1990 and an organizer of the first celebration 20 years ago.

"If we consider the environmental consequences of our purchases, our options in transportation, our investments, and our votes, we can send a powerful message to our public policymakers and our business leaders to do the same," Hayes emphasized.

Two years in the planning, Earth Day 1990 was orchestrated by a coalition of organizations—Greenpeace, the National Resources Defense Council, the Rainbow Coalition, the National Education Association, the Sierra Club, Planned Parenthood and many others—to evoke the idealism of Earth Day 1970, which launched the modern environmental movement.

Nevertheless, there was a widespread sense that this Earth Day had been sold to the American people with the same techniques employed by peddlers of soap and patriotism, but many also said that that was not a bad thing, given the urgency of the environmental cause.

"It feels mass-marketed," said Armando Martis, a Brooklyn dancer who was headed for Earth Day events in midtown Manhattan. "But that's good because it gets the message out to a lot of people."

Nearby, Rita Robbins said: "I don't think this is too commercial, although about 12 people have tried to sell me buttons—none of them biodegradable."

"It's a lot of hype, but for a good thing," said Donald Braler, a 30-year-old New York film technician.

There were other grumblings about companies with poor-to-disastrous pollution records that for weeks had run Earth Day advertisements expressing concern for the ecology, all without the blessing of Earth Day organizers, who said corporate America had wrapped itself in a green flag.

About 2,000 people were at the Washington Monument for Earthfest '90, an event sponsored by the Federal Environmental Protection Agency. John Denver sang a song he had written for the event, "The Flower That Shattered the Stone," and the Environmental Protection Agency administrator, William K. Reilly, gave a speech about the meaning of Earth Day.

Afterward, he told reporters: "It seems to me that the president has made a really fast start on the environment. I can't think of any 15-month period that produced as much by way of environmental initiatives, decisions, and participation by a president." Later, he sat alone backstage, sipping a soda as the reporters streamed past him in pursuit of the actor Tom Cruise.

In Central Park, a celebrity-studded

concert entertained the crowd in the afternoon, and the speakers included Governor Mario M. Cuomo, Mayor David N. Dinkins, the union activist Caesar Chavez, and others who talked of ecological responsibility.

But there were other settings in the park, less noticed, that gave meaning to the day: contrasts of young and old, of death and renewal—last year's faded grass beside a blaze of yellow daffodils; old people on benches, faces upturned to a warming sun, dreaming of long ago; and out on the lake, far from the exhaust of automobiles, a young boatman upright on the prow, floating like an ancient Chinese dragon.

On the Avenue of the Americas, huge throngs of people, at times all but immobilized by their own numbers, sampled a potpourri of arts and crafts, T-shirts and food, blaring rock music and exhortations of environmental responsibility.

Much of it had the flavor of a 1960s antiwar festival: Amid books on yoga and meditation were booths draped with handmade jewelry, yogis orange-robed and cross-legged, people strolling hand-in-hand, and vendors of vegetarian cuisine like mango nectar and "Southern fried cauliflower."

Recycling by the Homeless

The avenue was in shade most of the day between walls of soaring skyscrapers that symbolized corporate America—Exxon, the New York Hilton, CBS, Rockefeller Center, Paine Webber, Manufacturers Hanover Trust. The buildings were closed and the avenue, normally choked with vehicles, teemed with life. A stand sold biodegradable bags, recycled denim rugs, and bumper stickers and signs that proclaimed: "Acid Rain Is a Bad Trip," "No Cholesterol," and "High Fiber."

There were oases of beauty: baskets of pink petunias, red fuchsias, orange and purple impatiens, white peace lilies, and banks of begonias, geraniums, and ranunculus. Nearby, a booth sold lovely bonsai: miniature windblown Japanese elms and junipers.

But it was big trees that preoccupied a man on a bandstand. "Every ton of recycled paper will save 17 trees that give off precious oxygen for our planet," he said. He spoke, too, of recycling, but only a few people were actually doing it, as they do every day—the homeless, picking their way through the crowd, collecting cans and bottles.

Despite the vast numbers of people and events, there was something strangely serene about Earth Day. It harkened back to the measured values of another age, when Americans raised a barn in an afternoon and took walks of an evening or gathered around a radio in a crisis to listen, not just to what they wanted to hear but to a familiar voice talking sense.

RESOURCES

A Cleaner Environment:
What to Buy

By Mark K. Solheim,
with Melynda Dovel Wilcox and Sarah Young

Alicea Chevrier buys biodegradable bags for her trash and biodegradable diapers for her 15-month-old daughter. She buys milk and water in plastic jugs and takes the empty containers to the recycling center. She refuses to buy dishwashing detergent in clear plastic bottles, eggs in polystyrene cartons, or anything in aerosol cans because her recycling center won't take the empties.

Despite her efforts, says the Oxford, Michigan, woman, "I still feel guilty abut how much I put back into the environment."

People like Chevrier used to be considered zealots, but they're becoming the mainstream as more and more consumers consider how their shopping habits affect the environment: Discarded packaging and disposable products are clogging our landfills, aerosols and other personal-care products are polluting the air, and detergents, paints, and pesticides are ending up in groundwater, lakes, and streams.

The message seems to be sinking in. According to a Gallup survey, more than 90 percent of consumers are willing to make a special effort to buy products from companies trying to protect the environment. More than 90 percent say they would sacrifice some convenience, such as disposability, in return for environmentally safer products or packaging. And nearly 90 percent are willing to pay more for them.

Manufacturers and retailers are responding to growing environmental consciousness of their customers. Wal-Mart has begun to highlight certain products with signs on the shelves explaining why they're better for the environment than some other choice. Pringles potato chips are packaged in recycled paper, for example, and Duraflame logs are made from 100 percent recycled materials. K Mart and other retailers are also working with manufacturers to develop products and packaging that are better for the environment.

Whether these are the first rumblings

of a green revolution in retailing and manufacturing or merely the signs of a marketing fad remains to be seen. Meanwhile, until more large supermarkets climb on the bandwagon, co-ops, catalogs, and health food stores will probably remain the best sources for such products.

Picking the Package

Buying a product or packaging that has already been recycled or that can be recycled and then recycling it is probably the consumer's single most effective weapon in the war against garbage. Scores of products are routinely packaged in recycled paperboard. They may display the recycling symbol, three arrows forming a circle. Among them are Cheerios, Kellogg's cereals, Lipton Cup-a-Soup, Arm & Hammer baking soda, and S.O.S. pads, to name a few.

Aluminum and glass containers may also have had a previous life, but there's usually no indication of that on the package. Every glass bottle or jar on the supermarket shelves contains at least 20 percent recycled glass. Many of the aluminum cans you buy are also made of recycled materials—in 1988, 55 percent of all aluminum cans were recycled by consumers.

Plastic packaging is at the top of many environmentalists' hit lists, but for many products, plastic is actually a sensible choice. It can replace heavier, thicker materials, such as glass, saving on transportation costs and reducing breakage. Plastic lends itself to tamper-resistant packaging for food and medicine. It preserves freshness better than other materials.

But too much plastic is undeniably bad news. For one thing, it is tough to recycle because it has to be separated by the type of resin used in its manufacture. For another, plastic resists degrading, so most of our plastic trash will be around for centuries.

So far, plastic amounts to only about 7 percent of what we throw away—by weight. But because of its high density when compacted, it constitutes about 25 to 30 percent of our garbage by volume, according to some accounts. In the face of legislation limiting the use of plastic packaging in some states, the industry has started to promote recycling. Some large plastic containers now display the recycling symbol surrounding a number, which is a voluntary coding system introduced by the Society of the Plastics Industry to help recyclers sort plastic containers by resin composition.

Only two types of plastics are currently being recycled in any quantity: polyethylene terephthalate (PET), used in many plastic soda bottles, for example (coded 1); and high-density polyethylene (HDPE), used in milk jugs and detergent containers (coded 2).

The code on a container does not mean it is made from recycled plastic. If you live in an area where plastic can be recycled, the code simply enables recyclers to separate the containers into resin types. For examples of plastic products that are recycled, see "Your Guide to the Green Marketplace" on page 213.

If you can't pick containers that can be recycled, the less packaging the better. Microwave foods, with multiple layers of plastic and cardboard, are among products environmentalists love to hate. Individually packaged foods are another.

Delusions of Degradability

Degradable plastic bags and diapers are the most visible and heavily promoted environment-friendly products on the market. But environmentalists aren't so sure how "friendly" they are.

Biodegradable plastic is often made of polyethylene mixed with particles of cornstarch or cellulose that allow bacteria to attack the plastic and break it down. Photodegradable plastics contain light-sensitive additives that cause the plastics to break apart in the presence of light. A standard polyethylene bag could take 300 to 400 years to break down; one made of 6 percent cornstarch should degrade completely in 3 to 6 years.

That sounds like the perfect solution to the plastics problem, but that bag would have to be lying at the side of the road or otherwise exposed to the elements to completely degrade. A more likely destination is a landfill, where sunlight, oxygen, and rain, which trigger biodegradability, probably won't reach the material.

A report on plastics by Environmental Action, a lobbying and education organization, outlines other possible problems: toxicity when the chemicals do break down; diversion of public attention and energy from plastics recycling; and the unknown effect of those millions of particles of degraded plastic dust blowing around the earth and entering the groundwater.

Some supermarkets offer biodegradable plastic bags in which to carry your groceries home. Whether or not the plastic bag is degradable, choose paper bags instead. They're made from a renewable resource—trees—and unlike most plastic bags, they can be recycled.* Even better: Use reusable cotton shopping bags, available from catalogs and health food stores.

Diapers: The Bottom Line

We throw away some 18 billion diapers a year, and most of them end up in already clogged landfills. Basically made of plastic, wood pulp, and polyacrylate granules (for superabsorbency), traditional disposable diapers are only partly degradable under the best of conditions. To top it off, their bundles of wastes pose a growing health hazard as they leach into the groundwater.

Enter degradable disposable diapers. Ontario-based Dafoe & Dafoe (P.O. Box 365, Brantford, Ont. N3T–5N3; (519) 759–5952) makes Nappies Ultra biodegradable disposables. Nappies contain polyacrylates, so they rival the absorbency of traditional disposable diapers. They're sold at Toys-R-Us for less than $9 for 48 medium diapers. Sales of biodegradable TenderCare diapers, made by RMED International, 5555 E. 71st St., Suite 8300, Tulsa, OK 74136; (800) 344–6379, doubled last year. But TenderCares are expensive, at $59 for 176 medium-size, or 240 newborn-size, diapers. And because they contain no superabsorbent chemicals, they leak faster than other disposables, according to some parents who use them.

Unfortunately, in a landfill, where tightly sealed layers of trash aren't exposed to oxygen and light, biodegradables won't

---

*Editor's Note: Environmentalists have yet to reach a consensus on whether paper is preferable to plastic. Although paper is a renewable resource, paper manufacture typically produces toxic waste.

degrade much faster than regular disposable diapers.

For the environment, cotton diapers are the best solution. "We're changing the world one diaper at a time," says Thane Maynard of Cincinnati, putting an ironic twist on one degradable disposable diaper maker's slogan. To avoid the chore of washing diapers, the Maynards use a diaper service. In fact, a service may actually be cheaper than using regular disposables. The cost of having a diaper service deliver 60 diapers a week is $10 to $13, depending on where you live. The cost for a year of diaper service at $12 a week comes to $624. The cost for the same number of disposables at $10.50 for a case of 44 is $745.

The Air and the Water

Any product that contains a solvent, such as alcohol (hair spray, deodorant, body lotion, shaving cream), or a propellant, such as propane (aerosols), contributes polluting volatile organic compounds (VOCs).

VOCs react with sunlight to form lower-atmosphere ozone, which is bad and should not be confused with upper-atmosphere ozone, which blocks harmful ultraviolet rays and is being depleted by such chemicals as chlorofluorocarbons (CFCs). Most CFCs now come from air conditioners and industrial sources. The Environmental Protection Agency banned CFCs as a propellant in aerosols in 1978.

Detergents and household cleaners often contain phosphates and hazardous chemicals that pollute our lakes, streams, and drinking water. Paints and solvents also pollute the air during use but pose an even bigger problem to aquatic life and our drinking water when disposed of improperly. "Your Guide to the Green Marketplace" on page 213 lists environment-friendly alternatives. (For tips on disposal of household hazardous waste, write for the booklet *Household Hazardous Waste: What You Should and Shouldn't Do* from the Water Pollution Control Federation, 601 Wythe St., Alexandria, VA 22314. It's free with a self-addressed, stamped envelope.)

Lawn and garden care products are another hazard when used in excess or disposed of improperly. If you're concerned about the chemicals being used on your lawn and can't find reassurance on the labels or from your lawn-care company, you might want to try lawn and garden products from companies that claim their products are safe for the environment. Ringer, 9959 Valley View Rd., Eden Prairie, MN 55344; (800) 654–1047, has a full line of "100 percent natural, environmentally safe" lawn and garden care products, including pesticide made from chrysanthemums and supplies of natural predators.

Another source is the Necessary catalog—Necessary Trading Co., P.O. Box 305, Drawer 613P, New Castle, VA 24127; (703) 864–5103; $2. It sells products like fake-apple fruit fly traps and earthworm compost to improve your soil. Necessary will also send you a free list of books about organic gardening.

Your Guide to the Green Marketplace

Many manufacturers hope they have an answer to the growing concerns of their customers about garbage and pollution: "environment-friendly" products. These are popping up on store shelves and in mail-order catalogs at a fast and furious pace. Do these products live up to their promise? Will buying them really make a difference?

This guide can help you decide.

213

PROBLEM
Plastic Trash Bags

Of the 3½ pounds of garbage each of us throws away each day, most is neatly tied up in plastic trash bags and sent to the landfill. The bags take up a lot of space and don't easily degrade, leaving their contents more or less intact for centuries.

BETTER CHOICE: Paper or recycled-plastic bags.

Trade-off: Paper bags usually don't have handles and can tear and leak.

Comments: Thirty-gallon paper bags large enough for yard clippings are available from Seventh Generation catalog, (800) 456–1177, for $12.95 plus shipping. Paper grocery bags work nicely for kitchen trash. Biodegradable plastic trash bags (including Hefty and Glad) probably aren't worth the extra money (see "A Cleaner Environment" on page 209), but at least one brand of biodegradable trash bag, Good Sense, is made from recycled plastic and is comparable to conventional plastic bags in price (about $4 for 20 bags).

PROBLEM
Disposable Diapers

The 18 billion disposable diapers we throw away each year create a double whammy: Most end up in clogged landfills, where they stick around for centuries, and their bundle of babies' wastes may pose a health hazard if the wastes leach into the groundwater.

BETTER CHOICE: Cotton diapers; if disposable diapers are a must, use extra-thin diapers.

Trade-off: Cotton diapers are less convenient than disposables and need to be changed more often to prevent diaper rash and leaking. Many day-care centers require disposable diapers.

Comments: Cotton diapers from a diaper service may be the least expensive choice, according to a study published by waste-management consultant Carl Lehrburger.
   Biodegradable diapers (including TenderCares, Nappies, and Bunnies) usually cost more than other disposables, and environmentalists question how quickly they actually degrade in a landfill—although they degrade faster if composted.

Some are less absorbent than traditional disposables and have to be changed more often. Of the nonbiodegradable disposables, the best for the environment are extra-thin diapers, which take up less landfill space than thick disposables.

PROBLEM
Paper and Paper Products

More than 40 percent of all municipal waste is paper, ranging from newspapers and office trash to paper cups and plates. Most of it goes to already clogged landfills.

CULPRIT: Office paper, computer paper, stationery, and envelopes.

BETTER CHOICE: Recycled paper. Examples: Earth Care Paper Inc., P.O. Box 3335, Madison, WI 53704; (608) 256–5522, offers a wide range of stationery, cards, office and computer paper, gift-wrapping paper, and envelopes; Conservatree Paper Co., Ten Lombard St., Suite 250, San Francisco, CA 94111; (800) 522–9200 or, in California, (415) 433–1000, sells computer, high-speed copier, and laser-printer paper,

stationery, and offset and text papers for printing. Recycled greeting cards are also available in stationery stores.

Trade-off: Recycled paper generally costs 10 to 30 percent more than paper made from virgin wood pulp. Recycled greeting cards are comparably priced.

Comments: Producing 1 ton of paper from discarded waste paper, according to Worldwatch Institute, uses about half the energy and half the water, results in 74 percent less air pollution and 35 percent less water pollution, saves 17 pulp trees and creates more jobs compared with producing a ton of paper from virgin pulp.

CULPRIT: Paper towels, napkins, facial tissue, and toilet paper.

BETTER CHOICE: Sponges, cloth towels and napkins, or recycled products. Examples: Seventh Generation catalog, (800) 456–1177, features a line of recycled paper towels, dinner napkins, facial tissue, and toilet paper. The Co-op America catalog, Ten Farrell St., South Burlington, VT 05403; (802) 658–5507 ($3.50 service charge to nonmembers), also offers toilet paper. The Canadian supermarket chain Loblaw includes recycled paper products in its "Green" line.

Trade-off: Higher cost. The toilet paper from Co-op America, for example, costs $27 for 48 rolls of 500 sheets, plus shipping. You can buy the same amount of nonrecycled toilet paper, which is softer, for $13.

CULPRIT: Coffee filters.

BETTER CHOICE: Eco-Filter, a reusable unbleached cotton coffee filter available for $5 or less, depending on size, from Earthen Joys catalog, 1412 11th St., Astoria, OR 97103; send $1 for the catalog.

Trade-off: Eco-Filter has to be washed, but it lasts over a year, so it's much less expensive than individual filters.

Comments: Coffee filters, like other chlorine-bleached paper products, may leave a residue of dioxin, a suspected carcinogen and environmental hazard. Natural Brew unbleached coffee filters from Rockline Inc., Box 1007, Sheboygan, WI 53082, are virtually dioxin-free; they're light brown instead of white, but they sometimes cost less than other national brands.

PROBLEM
Energy-Inefficient Light Bulbs

The typical household spends $68 a year in electricity costs for lighting. Switching from incandescent bulbs to more efficient lighting would save that household about $41 a year.

BETTER CHOICE: Compact fluorescent bulbs, such as those sold by General Electric, Sylvania, and Philips (cost is about $8

to $20 per bulb—which may include a reusable ballast that screws into the light fixture), or Capsylite light bulbs by Sylvania ($2 to $3.50 per bulb), which are superefficient incandescent bulbs.

Trade-off: Compact fluorescents are initially much more expensive than incandescent bulbs (which cost about 70 cents each), but they last longer and save money over the long term.

Comments: Fluorescent bulbs last ten times as long as incandescents; Capsylite bulbs last four times as long. Energy-efficient bulbs may be hard to find—check home-supply or lighting stores.

PROBLEM
Packaging

Nearly a third of what we throw away is empty containers and packages. Most of our trash doesn't get recycled; much of it—particularly plastic—isn't easily recyclable; and even paper doesn't degrade much in landfills. Result: We're running out of landfill space.

CULPRIT: Paperboard packaging.

BETTER CHOICE: Recycled paperboard packaging.

Trade-off: None.

Comments: A slew of products are packaged in recycled paperboard, from Kellogg's

Corn Flakes to Crayola Crayons. Most display the recycling symbol—three arrows forming a circle. Note that the "recyclable" symbol is similar to the "recycled" symbol. Read the description underneath the symbol to be sure.

CULPRIT: Plastic milk and beverage containers.

BETTER CHOICE: Cardboard or glass containers.

Trade-off: Cardboard milk cartons may leach dioxin, a suspected carcinogen and environmental hazard that is a by-product of the bleaching process; glass bottles are heavier and are breakable.

Comments: Wax-covered cardboard generally isn't recyclable, but it's degradable; glass is recyclable. If plastic can be recycled in your area, plastic milk and beverage containers are the best choice. (For more information about plastics recycling, see "A Cleaner Environment" on page 209.)

CULPRIT: Multimaterial juice packs, such as those made by Hi-C or Ocean Spray.

BETTER CHOICE: Juice made from concentrate or juice packs of recyclable tin cans.

Trade-off: Convenience; no straw included in juice packs.

Comments: Single-serving juice containers made of foil, paper, and plastics are not recyclable.

CULPRIT: Plastic containers of liquid washing-machine detergent and household cleaners.

BETTER CHOICE: Powdered detergents

and cleaners in paperboard packaging, especially concentrated or superconcentrated formulas; liquid detergent in recycled plastic containers. Example: Spic-and-Span Pine, Tide, and Cheer containers available in the Northwest use some recycled plastic.

Trade-off: Many powdered detergents contain more phosphates than liquids; phosphates are a water pollutant (see "Detergents and Household Cleaners" on page 219).

Comments: Because of lack of markets for recycled plastic, recycled plastic containers may not be recycled a second time.

CULPRIT: Plastic dishwashing detergent containers.

BETTER CHOICE: Recyclable or recycled plastic containers; refills. Examples: Colgate-Palmolive is testing Palmolive dishwashing detergent in soft plastic pouches, called doy packs, to be used with refillable containers; Dial Corp. is introducing Purex in "bag-in-box" packaging, a collapsible plastic bottle.

Trade-off: Doy packs lack rigidity; pouring refills into reusable containers is inconvenient.

Comments: One reason Colgate-Palmolive is testing doy packs is that Palmolive currently comes in clear PET plastic bottles, which aren't recyclable, in many areas. Most plastics recyclers will accept PET soda bottles and plastic containers made of HDPE.

CULPRIT: Plastic fabric softener containers.

BETTER CHOICE: Concentrated fabric

softener. Examples: Procter & Gamble is test-marketing a Downy fabric softener concentrate in a paperboard refill carton in the Baltimore-Washington area, for a cost about 10 percent less than the Downy in a plastic container. Amway, (800) 544–7167 for nearest distributor, sells a concentrated fabric softener—32 ounces of its softener equals 216 ounces of Downy, according to the company; cost is $6.25, versus $9.34 for the same amount of Downy.

Trade-off: Convenience.

CULPRIT: Soft-soap dispensers

BETTER CHOICE: Bar soap, which is also less expensive; Colgate-Palmolive is marketing nationally a Softsoap "cotelle pack"—a squeezable, rectangular refill pouch.

Trade-off: Convenience.

CULPRIT: Toothpaste pumps.

BETTER CHOICE: Tubes of toothpaste, which use less material, take up very little landfill space, and are less expensive.

Trade-off: Convenience: such products may be difficult to find.

CULPRIT: "Blister" packaging, in which the product is sandwiched between plastic and cardboard.

BETTER CHOICE: The same product without the extra packaging.

Trade-off: Convenience; such products may be difficult to find.

CULPRIT: Polystyrene cups and plates.

BETTER CHOICE: Paper cups and plates or washable, reusable cups and plates.

Trade-off: Polystyrene is lightweight and is an excellent insulator for hot and cold food; paper cups sometimes leak.

Comments: Polystyrene products have been banned in some communities because they don't degrade easily and generally are not recycled. Some polystyrene is manufactured using ozone-depleting chlorofluorocarbons (CFCs), but the foam food-service packaging industry began using a much less destructive form of CFCs in 1988. A recently formed company is beginning to set up polystyrene recycling plants, and some McDonald's restaurants in the Northeast collect polystyrene containers for recycling.

CULPRIT: Polystyrene egg cartons.

BETTER CHOICE: Cardboard egg cartons, which are recyclable and even compostable. Trio Products makes clear plastic egg cartons from recycled beverage bottles, which are available in some Kroger stores and other chains under Tyson, Holly Farms, and other brand names.

Trade-off: Eggs in recycled clear-plastic cartons can cost 10 to 20 cents more per dozen.

CULPRIT: Add-water-and-shake pancake and waffle mixes with throwaway plastic shakers, such as Bisquick's Shake'n Pour and Aunt Jemima Pancake Express.

BETTER CHOICE: Pancakes and waffles made from bulk-packaged mix.

Trade-off: Convenience, but you pay up to four times less for bulk mix.

PROBLEM
Personal-Care Products

Deodorants, hair spray, and other aerosol personal-care products contain propellants (usually isobutane, butane, and propane) and solvents (including ethanol, isopropyl alcohol, and propylene glycol) that are volatile organic compounds (VOCs). VOCs, which also come from car exhaust, interact with sunlight and other compounds to form lower-atmosphere ozone, a pollutant and "greenhouse" gas that contributes to global warming. Personal-care products that come in pumps or even lotions may also have a high solvent content. Solvents can get into lakes and streams and pose a hazard to aquatic life and are a suspected groundwater contaminant.

CULPRIT: Alcohol-based hair spray.

BETTER CHOICE: Cosmosol Ltd. will begin marketing a water-based hair spray called New Idea, which will cost $3.25 for 8 ounces and should be available by the end of February [1990].

Trade-off: None.

Comments: In California, where VOC-emitting consumer products will be regulated as of 1992, solvents in hair spray account for some 27 tons of emissions per day, the second largest contributor of VOCs

in consumer products; aerosol paints are the first. Pumps tend to contain even more alcohol than aerosols, but because they have better "application efficiency," they tend to release fewer VOCs into the atmosphere than aerosols. Other hair-care products, such as mousse and even shampoo, contribute VOCs.

CULPRIT: Aerosol underarm deodorants.

BETTER CHOICE: Sticks or roll-ons.

Trade-off: Roll-ons are slower drying than aerosols.

Comments: Le Crystal, a body deodorant by Nature de France, is free of perfumes and chemicals and can last a year or more. Available for $15 from the Ecology Box, 425 E. Washington, No. 202, Ann Arbor, MI 48104; (800) 634–6715 or, outside Michigan, (313) 662–9131 (call collect).

PROBLEM
Detergents and Household Cleaners

Detergents and household cleaners often contain phosphates, surfactants (foaming and penetrating agents), organic solvents, and toxins that can enter lakes and streams. They are hazardous to wildlife and can end up in drinking water. Many powdered detergents, particularly automatic dishwasher detergents, contain high levels of phosphates. Liquid detergents may also contain phosphates. A number of states limit phosphate content, and many manufacturers voluntarily limit it. All detergents also contain surfactants, which may not be biodegradable. Many household cleaners contain solvents, which are VOCs and contribute to low-level ozone, and toxic substances. Aerosols especially contribute airborne VOCs.

CULPRIT: Phosphate-containing, non-biodegradable dishwashing and laundry detergents and household cleaners.

BETTER CHOICE: Phosphate-free, biodegradable detergents. You can get an entire line of such detergents and cleaning products from Amway, (800) 544–7167 for nearest distributor; Ecover, available through Seventh Generation catalog, (800) 456–1177; and Winter White, available from Mountain Fresh Products, P.O. Box 40516, Grand Junction, CO 81504; (303) 243–8835).

Trade-off: Some limited-distribution products are higher priced.

CULPRIT: Air fresheners.

BETTER CHOICE: An open box of baking soda or cedar blocks.

Trade-off: Convenience.

Comments: Aerosol air fresheners have a VOC content of 90 to 99 percent. Effective February 28, 1990, New Jersey will limit the solvent content of air fresheners to 50 percent by weight. Solid air fresheners have a VOC content of 5 to 10 percent.

CULPRIT: Drain cleaners.

BETTER CHOICE: Plunger, snake, or other mechanical means for opening drain.

Trade-off: Convenience; possibly less effective.

Comments: A company named Bio-Care recently began offering the food and restaurant industry a "biotechnological" drain-care service—bacteria that "eat" grease that accumulates in pipes. The service is expected to be available to consumers by the end of the year. The company offers a similar product to consumers for use in a septic tank. Call (800) 421–9740 or, in California, (800) 468–5666 for more information.

CULPRIT: Glass cleaners.

BETTER CHOICE: Two tablespoons of vinegar to 1 quart of water, or just plain water with a squeegee.

Trade-off: Convenience; slightly less effective.

Comments: Pumps and concentrates are preferable to aerosols.

PROBLEM
Paints and Solvents

Oil-based paints, paint solvents and thinners, paint removers, wood stains and preservatives, and waxes all contain solvents that pose environmental risks when disposed of improperly and health risks during and after use. Solvents contribute to low-level ozone, an air pollution hazard; solvents and other toxins that get into lakes and streams are also hazardous to aquatic life and are a suspected groundwater contaminant.

CULPRIT: Aerosol paints.
BETTER CHOICE: Brushable paints.

Trade-off: Brushable paints are less convenient but also less expensive.

Comments: Aerosol paints are the number-one contributor of VOCs in consumer products in California.

CULPRIT: Oil-based paints, paint solvents, and thinners.

BETTER CHOICE: Latex paints, which have fewer VOCs and can be cleaned up with soap and water.

Trade-off: Latex paints often aren't as durable as oil-based paints, particularly outside.

Comments: Latex paints, although preferable, are not perfect. They do contain low levels of hazardous additives, such as zinc. At least two companies sell "plant chemistry" nonsynthetic paints, paint removers, wood finishes, and waxes: Auro Organic Paints, imported from Germany by Sinan Co., Natural Building Materials, P.O. Box 857, Davis, CA 95617; (707) 427–2325; and Livos PlantChemistry, 2641 Cerrillos Rd., Santa Fe, NM 87501; (505) 988–9111. The cost of 0.75 liters (about three-fourths of a quart) of Livos white paint, for example, is $5.85, compared with $5.55 for a quart of Duron interior latex paint.

Parts of this listing are from the *Conservation Directory* of the National Wildlife Federation. Reprinted by permission.

American Rivers
801 Pennsylvania Ave. SE, Suite 303
Washington, DC 20003
(202) 547-6900

Works to protect wild, natural, and free-flowing rivers and their landscapes and acts as an information resource for activists. Works on legislation and assists local and state organizations in conservation projects. Publishes newsletter to update members on river-related legislation action, provides contact names, and lists recreational opportunities.

Center for Marine Conservation
1725 DeSales St. NW
Washington, DC 20036
(202) 429-5609

A nonprofit membership organization (formerly the Center for Environmental Education) dedicated to protecting marine wildlife and habitats and to conserving coastal and ocean resources. Supports major international efforts to protect species threatened by international trade. Conducts policy research, promotes public awareness, involves citizens in public policy decisions, and supports domestic and international conservation programs for marine species and their habitiats. Members: More than 110,000 worldwide.

Chesapeake Bay Foundation
162 Prince George St.
Annapolis, MD 21401
(301) 268-8816 (Annapolis)
(301) 269-0481 (Baltimore)
(301) 261-2350 (Washington)

A nonprofit conservation organization supported by charitable contributions from members, philanthropic foundations, and corporations. CBF's goal is to promote and contribute to the orderly management of the Chesapeake Bay with a special emphasis on maintaining a level of water quality that will support the bay's diverse aquatic species. Programs include in-the-field instruction in estuarine ecology, scientific investigation, and legal representation concerning conservation and management of estuarine resources, and the preservation and management of significant bay lands. Preserves and manages more than 3,500 acres of bay land. Members: More than 70,000. Dues: $20 annually for individual; $30 for family. Members receive four issues of the *CBF News* and an annual report.

Citizens Clearinghouse for Hazardous Waste
Box 926
Arlington, VA 22216
(703) 276-7070

Provides assistance to citizens and grass-roots groups fighting hazardous waste problems. Communicates information through a variety of environmental publications, offers workshops to help communities organize and solve environmental problems, and provides individuals, groups, and communities with scientific and technical assistance in interpreting and using technical information.

Clean Water Action Project
317 Pennsylvania Ave. SE
Washington, DC 20003
(202) 547-1196

Works for clean, safe, affordable water, control of toxic chemicals, and the protection of natural resources. Provides technical and organizing assistance to groups involved in fighting incinerators, cleaning up dumps, and protecting groundwater. Also provides strategy and support to citizens' efforts to inform the public on electoral candidates' environmental positions.

The Cousteau Society
930 West 21st St.
Norfolk, VA 23517
(804) 627-1144

An international, nonprofit, membership-supported organization dedicated to the protection and improvement of the quality of life, with an emphasis on the marine environment. Current focus: Informing and alerting the public through television/ film, research, lectures, books, and publications focusing on humanity's interaction with the environment. Members: 320,000. Volunteer programs: Many volunteers work in the Norfolk office.

Defenders of Wildlife
1244 19th St. NW
Washington, DC 20036
(202) 659-9510

A national nonprofit organization whose goal is to preserve, enhance, and protect wildlife and preserve the integrity of natural ecosystems. Current focus: Protecting and restoring habitats and wildlife and promoting wildlife appreciation and education. Members: 80,000-plus members and supporters. Volunteer programs: An activist network consisting of more than 6,000 individuals.

Ducks Unlimited
One Waterfowl Way
Long Grove, IL 60047
(708) 438-4300

Ducks Unlimited raises money for developing, preserving, restoring, and maintaining waterfowl habitat in North America and educates the public concerning wetlands and waterfowl management. Members: More than 500,000. Volunteer programs: Has almost 4,000 volunteer committees in all parts of the U.S., Canada, and Mexico. Contact the above address for regional information.

Earth Island Institute
300 Broadway, Suite 28
San Francisco, CA 94133
(415) 788-3666

The Earth Island Institute initiates and supports internationally oriented action projects for the protection and restoration of the environment. Current focus: Stopping the killing of dolphins, and fostering environmental protection and restoration in Central America. Members: 30,000. Members receive the quarterly international newsmagazine, *Earth Island Journal*. Volunteer programs: Program internships are available in each project area, as are journalism internships.

Environmental Action/Environmental Action Foundation
1525 New Hampshire Ave. N.W.
Washington, DC 20036
(202) 745-4870

EA, a national nonprofit membership organization, works for enactment of the strongest possible environmental laws and publishes the bimonthly *Environmental Action* magazine. Its affiliated political action committee, EnAct/PAC, endorses pro-environment candidates and spotlights the worst members of Congress through the "Dirty Dozen" campaign. The Environmental Action Foundation helps citizens organize; assists grass-roots groups; litigates to protect consumers from abuse by electric utilities; and conducts public education and research about toxics, pesticides, recycling and solid wastes, global warming, energy conservation, and nuclear weapons and power-generation facilities.

Environmental Defense Fund, Inc.
257 Park South
New York, NY 10010
(212) 505-2100

The EDF, with headquarters in New York and offices nationwide, is a conservation organization whose work spans global issues such as ocean pollution, rainforest destruction, and the greenhouse effect. Since its founding in 1967 in the effort to save the osprey and other wildlife from DDT, it has used teams of scientists, economists, and attorneys to develop economically viable solutions to environmental problems. Members: More than 150,000.

Friends of the Earth
218 D St. SE
Washington, DC 20003
(202) 544-2600

Dedicated to the conservation, protection, and rational use of the earth. Focus is on saving the ozone layer, ending tropical deforestation, fighting global warming, tackling the waste crisis, protecting the oceans, encouraging corporate responsibility for the environment, ending nuclear weapons production, and redirecting tax dollars to the environment. Publishes several periodicals. Affiliated with 37 FOE groups around the world. Recently merged with the Oceanic Society and the Environmental Policy Institute. Members and supporters: 50,000.

Global Tomorrow Coalition
1325 G St. NW, Suite 915
Washington, DC 20005-3104
(202) 628-4016

Organizes task forces on community and national levels to discuss environmental issues and work toward passing legislation affecting the future of the world's natural resources. Addresses global trends in population growth, wasteful resource consumption, environmental degradation, and sustainable development.

Greenpeace, USA, Inc.
1436 U St. NW
Washington, DC 20009
(202) 462-1177

Greenpeace is an international organization dedicated to protecting the natural environment. Works to shield the environment from nuclear and toxic pollution, halt the slaughter of whales and seals, stop nuclear weapons testing, and curtail the arms race at sea. Also campaigns against the mining and reprocessing of nuclear fuel, the exploitation of Antarctica, and the destruction of marine resources through indiscriminant fishing. Members: 4.5 million supporters worldwide, offices in 23 countries.

International Marinelife Alliance-USA
94 Station St., Suite 645
Hingham, MA 02043
(617) 383-1209

An international nonprofit membership organization committed to supporting
practical and positive measures to protect
and restore the health and diversity of marine environments, with emphasis on coral
reef ecosystems (especially in developing
countries). Works through research, education, and economic and technical cooperation. Involved in offshore oil regulations
and the development of a national marine
disaster response plan that will assess and
implement cleanup and rehabilitation of
wildlife damaged by oil spills.

Izaak Walton League of America
1401 Wilson Blvd., Level B
Arlington, VA 22209
(703) 528-1818

A national conservation organization
started in 1922 that works toward the protection of America's land, water, and air resources. Current focus: Acid rain, clean
water, stream protection, Chesapeake Bay
cleanup, outdoor ethics, public land management, soil erosion, clean air, and water-
fowl/wildlife protection. Members: 50,000
(400 chapters). Volunteer programs: Save
Our Streams, Wetlands Watch, Oregon
Public Lands Restoration Task Force.

National Audubon Society
950 Third Ave.
New York, NY 10022
(212) 832-3200

Conserves plants and animals and their
habitats; promotes rational strategies for energy development and use, stressing conservation and renewable resources; protects
life from pollution, radiation, and toxic substances; furthers the wise use of land and
water; seeks solutions for global problems
involving the interaction of population, resources, and the environment. Members:
600,000. Volunteer Programs: Annual
Christmas Bird Count and Breeding Bird
Census; Audubon Activist Network; internships; Citizen's Acid Rain Monitoring Network; involvement through local chapters,
state and regional offices.

National Coalition against the Misuse of
Pesticides
530 7th St. SE
Washington, DC 20003
(202) 543-5450

A nonprofit membership organization
founded ten years ago to serve as a national
network committed to pesticide safety and
the adoption of alternative pest-
management strategies that reduce or eliminate a dependency on toxic chemicals. Provides useful information on pesticides and
alternatives to their use; publishes newsletters *Pesticides and You* and *NCAMP's Technical
Report*.

National Toxics Campaign
1168 Commonwealth Ave.
Boston, MA 02134
(617) 232-4014

The National Toxics Campaign works
with more than 1,400 grass-roots environmental groups nationwide to assist them in
defending communities against toxic chemical threats. Provides organizing and technical assistance, media help, educational
and training materials, and workshops.
Runs the nation's only full-service lab that
provides low-cost, reliable testing to communities at risk from toxic contamination.
Publishes quarterly magazine. Annual
membership fee: $15 for individuals, $100
for groups.

National Wildlife Federation
1400 16th St. NW
Washington, DC 20036-2266
(202) 797-6800

Promotes the wise use of natural resources and protection of the global environment. Distributes periodicals and educational materials, sponsors outdoor education programs in conservation, and litigates environmental disputes in an effort to conserve natural resources and wildlife. Current focus: Forests, energy, toxic pollution, environmental quality, fisheries and wildlife, wetlands, water resources, public lands. Members and supporters: More than 5 million. Volunteer programs: The National Wildlife Federation includes 52 state and territorial affiliates, all of which rely on volunteer support.

Natural Resources Defense Council
40 West 20th St.
New York, NY 10011
(212) 727-2700

Seeks to protect natural resources and improve the quality of the human environment. Combines legal action, scientific research, and citizen education in a highly effective environmental protection program. Major accomplishments have been in the area of energy policy and nuclear safety, air and water pollution, urban environmental issues, toxic substances, and natural resources conservation. Members: 95,000.

The Nature Conservancy
1815 North Lynn St.
Arlington, VA 22209
(703) 841-5300

An international nonprofit membership organization committed to preserving biological diversity by protecting natural lands and the life they harbor. Cooperates with educational institutions, public and private conservation agencies. Works with states through "natural heritage programs" to identify ecologically significant natural areas. Manages a system of over 1,500 nature sanctuaries nationwide. Works with Central and South American conservation organizations to identify and protect natural lands and tropical rainforests.

Ocean Alliance
Suite 2, Bldg. E
Fort Mason Center
San Francisco, CA 94123
(415) 441-5970

A nonprofit organization working to preserve sea life and ocean habitat through conservation and education. Oversees conservation projects, including opposing commercial whaling, preventing the drowning of marine mammals in nets, and establishing new national marine sanctuaries. Promotes education through school programs such as Project Ocean, The Whale Bus, and Adopt-a-Beach. Publishes the quarterly *Ocean Ally* and maintains a library. Members: 3,500. Volunteer programs: Volunteers are invited to help on conservation projects and in office clerical positions.

Rainforest Action Network
301 Broadway, Suite A
San Francisco, CA 94133
(415) 398-4404

A nonprofit activist organization working to save the world's rainforests. Begun in 1984, it works internationally in cooperation with other environmental and human rights organizations on campaigns to protect rainforests. Provides financial support for groups in tropical countries to preserve forest lands. Methods include direct action, grass-roots organizing, and media outreach. Current focus: Tropical timber, indigenous peoples, and multilateral development banks. Members: 30,000. Volunteer programs: Internships are available.

Renew America
1400 16th St. NW
Suite 710
Washington, DC 20036
(202) 232-2252

A nonprofit environmental organization dedicated to the development of a safe and sustainable society. Oversees two national programs. Searching for Success collects and evaluates data on working environmental programs and makes the information available to communities, government agencies, and businesses. The State of the States program compares and ranks state environmental efforts.

Sierra Club
730 Polk St.
San Francisco, CA 94109
(415) 776-2211

Promotes conservation of the natural environment by influencing public policy; practices and promotes the responsible use of the earth's ecosystems and resources; educates and enlists humanity to protect and restore the quality of the natural and human environment. Current focus: Clean Air Act reauthorization, Arctic National Wildlife refuge protection, Bureau of Land Management wilderness/desert national parks protection, toxic waste regulations, global warming/greenhouse effect. Members: 550,000 (57 chapters). Volunteer programs: Extensive opportunities available throughout the country in conservation campaigns and outdoor activities.

Trout Unlimited
501 Church St. NE
Vienna, VA 22180
(703) 281-1100

A nonprofit international organization dedicated to the protection of clean waters and the enhancement of trout, salmon, and steelhead habitat. The national office works with Congress and federal agencies for the protection and wise management of America's fishing waters, sponsors seminars, funds research projects, and administers the nationwide network of chapters. Members: 68,000 (495 chapters, with affiliates in eight other countries). Volunteer programs: All chapters welcome the assistance of volunteers.

Union of Concerned Scientists
26 Church St.
Cambridge, MA 02238
(617) 547-5552

A nonprofit organization of scientists and other citizens concerned about the impact of advanced technology on society. Advocates energy strategies that minimize risks to public health and safety, provide for efficient and cost-effective use of energy resources, and minimize damage to the global environment. Also advocates defense policies and negotiated arms agreements that reduce the risk of nuclear war, benefit U.S. security interests, and enhance the nation's economic strength. Sponsorship: Over 100,000.

U.S. Public Interest Research Group
215 Pennsylvania Ave. SE
Washington, DC 20003
(202) 546-9707

A nonprofit, nonpartisan environmental and consumer advocacy organization, founded in 1983, which represents the public's interest in the areas of consumer and environmental protection, energy policy, and governmental and corporate reform. Current focus: clean air, pesticide and product safety, global warming and ozone depletion, and reducing the use of toxics. Members: 1 million.

The Wilderness Society
900 17th St. NW
Washington, DC 20006
(202) 833-2300

A nonprofit membership organization, founded in 1935, devoted to preserving wilderness and wildlife, protecting America's forests, parks, rivers, and shorelands. Members: More than 300,000. Welcomes membership inquiries, contributions, and bequests.

World Resources Institute
1709 New York Ave. NW, 7th fl.
Washington, DC 20006
(202) 638-6300

A policy research center that helps governments, the private sector, and environmental and development organizations address the fundamental question of meeting human needs and nurturing economic growth while preserving natural resources and environmental integrity. Current focus: Tropical forests, biological diversity, sustainable agriculture, energy, climate change, atmospheric pollution, benign environmental technologies, international institutions, and economic incentives for sustainable development.

World Wildlife Fund and the Conservation Foundation
1250 24th St. NW
Washington, DC 20037
(202) 293-4800

A private U.S. organization working worldwide to protect endangered wildlife and wildlands. Its top priority is conservation of tropical forests. Current focus: A campaign to save the African elephant, sustainable development programs in Asia and Africa, the panda management plan with China TRAFFIC—a program that monitors trade in wild plants and animals—and a jaguar preserve in Belize. Affiliated with the international WWF network, which includes national organizations and associates in 26 countries across 5 continents. Membership: More than 600,000.

Zero Population Growth
1400 16th St. NW
Suite 320
Washington, DC 20036
(202) 332-2200

A nonprofit membership organization that works to achieve a sustainable balance between the earth's population, its environment, and its resources. Primary activities include publishing newsletters and research reports, developing in-school population education programs, and coordinating local and national citizen action efforts. Members: 30,000. Volunteer programs: Legislative Alert Network (congressional support for critical population/environmental legislation); Media Targets (volunteers write to media to encourage better reporting); Teacher's Pet Project (volunteers help with teacher-training workshops).

INDEX

Note: Page references in *italic* indicate photographs.

Recycling, 5–8, 18–25
  of aluminum, 6, 19
  of automobile batteries, 117
  of automotive oil, 162
  Earthworm, Inc. and, 18–19
  of glass, 6
  "Lucky Can" machine and, 24–25
  of paper, 3, 6–7
  in Perkasie, Pennsylvania, 13–17
  of plastics, 6
  of polystyrene, 4
  of yard waste, 6
Red Wing, Minnesota, pleasure boating and,
  150–51
Refrigerators, CFC emissions and, 68–71
Renew America, 227
Rhinoceros horn, ban on sales of, 176
Rivers and Harbors Act, 121
Rocky Flats, Colorado, pollution in, 120, 121
Rodale Research Center, Pennsylvania, *82*

S

Santa Ana, Costa Rica, *xix*
Sauget, Illinois, water pollution in, 154–56
Schwarzschild, Bert, songbird hunting and,
  189–91
Sea turtles, on Cyprus, 59–61
Senegal, Africa, *xvi*
Shields, Dale, pelican protection and, 194–95,
  195 Sierra Club, 227
Smog, *62*, 65
Solar panels, *26*
Solar power
  automobiles and, 37–38
  electricity from, 34–38
Solid waste. *See also* Landfills
  disposal of
      in Antarctica, 99–101, 108–9, *109*
      in United States, 1–17
  ocean littering and, 57–59
  source reduction of, 8, 9–10
Solvents
  air pollution by, 66

alternatives to, 220–21
Songbirds, hunting of, 189–91
South Africa, elephant poaching in, 176
South Pole. *See also* Antarctica
  exploration of, 96–97
Stoves, fuel-efficient, 143–44, *144*
Stratospheric ozone, 63. *See also* Ozone
  international commitments on, 67
  protection of, 66–67
  thinning of, 64, 94–95, 197
Subarctic grasslands, greenhouse effect and, 29
Summer heat, greenhouse effect and, 30
Sweetlips coral fish, *46*

T

Terrebonne Parish, Louisiana, 160, 161
3M Company, Minnesota
  environmental awareness of, 123–24
  pollution by, 126–27
Tin cans, recycling of, 24–25
Toxaphene, ban on, 156
Toxic waste, 111–12, 114–15
  corporate environmental awareness and,
      123–27
  corporate pollution and, 118–19, 120–22
  Love Canal and, 115
  research on, 112–13
  storage of, 151–52
  Whately, Massachusetts and, 116–17
Trans-Alaska Pipeline, impact of, 104–5
Trash museum, New Jersey, 11–13, *12*
Trees
  as crops, 45
  greenhouse effect and, 42–45
Tropical forests, *128*. *See also* Amazon basin
  beef production and, 146
  in Brazil, destruction of, 43, 129–38
  economic value of, 178
  environmental protests and, 131–32
  fragility of, 132
  Grande Carajas Program and, 133
  greenhouse effect and, 42, 130, 131
  hydroelectric dams and, 147
  life forms of, 131